Plant Science Research and Practices

# *Acacia*

## Characteristics, Distribution and Uses

# PLANT SCIENCE RESEARCH AND PRACTICES

Additional books and e-books in this series can be found on Nova's website under the Series tab.

PLANT SCIENCE RESEARCH AND PRACTICES

# *ACACIA*

## CHARACTERISTICS, DISTRIBUTION AND USES

AIDE MATHESON
EDITOR

Copyright © 2018 by Nova Science Publishers, Inc.

**All rights reserved.** No part of this book may be reproduced, stored in a retrieval system or transmitted in any form or by any means: electronic, electrostatic, magnetic, tape, mechanical photocopying, recording or otherwise without the written permission of the Publisher.

We have partnered with Copyright Clearance Center to make it easy for you to obtain permissions to reuse content from this publication. Simply navigate to this publication's page on Nova's website and locate the "Get Permission" button below the title description. This button is linked directly to the title's permission page on copyright.com. Alternatively, you can visit copyright.com and search by title, ISBN, or ISSN.

For further questions about using the service on copyright.com, please contact:
Copyright Clearance Center
Phone: +1-(978) 750-8400        Fax: +1-(978) 750-4470        E-mail: info@copyright.com.

### NOTICE TO THE READER

The Publisher has taken reasonable care in the preparation of this book, but makes no expressed or implied warranty of any kind and assumes no responsibility for any errors or omissions. No liability is assumed for incidental or consequential damages in connection with or arising out of information contained in this book. The Publisher shall not be liable for any special, consequential, or exemplary damages resulting, in whole or in part, from the readers' use of, or reliance upon, this material. Any parts of this book based on government reports are so indicated and copyright is claimed for those parts to the extent applicable to compilations of such works.

Independent verification should be sought for any data, advice or recommendations contained in this book. In addition, no responsibility is assumed by the publisher for any injury and/or damage to persons or property arising from any methods, products, instructions, ideas or otherwise contained in this publication.

This publication is designed to provide accurate and authoritative information with regard to the subject matter covered herein. It is sold with the clear understanding that the Publisher is not engaged in rendering legal or any other professional services. If legal or any other expert assistance is required, the services of a competent person should be sought. FROM A DECLARATION OF PARTICIPANTS JOINTLY ADOPTED BY A COMMITTEE OF THE AMERICAN BAR ASSOCIATION AND A COMMITTEE OF PUBLISHERS.

Additional color graphics may be available in the e-book version of this book.

## Library of Congress Cataloging-in-Publication Data

ISBN: 978-1-53614-237-2

*Published by Nova Science Publishers, Inc. † New York*

# CONTENTS

**Preface** vii

**Chapter 1** Stem and Wood Characterization of *Acacia melanoxylon* as an Introduced Species in Europe 1
*Ofélia Anjos, António J. A. Santos, Isabel Miranda, Sofia Knapic and Helena Pereira*

**Chapter 2** Ecology of Acacieae 27
*Marcos S. Karlin and Ulf O. Karlin*

**Chapter 3** Regeneration in *Acacia Mangium* Willd. Plantation in Mt. Makiling Forest Reserve, Philippines: Implications to Restoration Ecology 83
*Marilyn S. Combalicer, Jonathan O. Hernandez, Pil Sun Park and Don Koo Lee*

**Chapter 4** NMR Spectroscopy in Solution of Gum Exudates Located in Venezuela 113
*Maritza Martínez and Gladys León de Pinto*

**Chapter 5** Reconstruction of Ancestral Character States in Neotropical Ant-*Acacias* 161
*Jessica Admin Córdoba de León and Sandra Luz Gómez Acevedo*

| | | |
|---|---|---|
| **Chapter 6** | *Acacia* Pollen: An Important Cause of Pollinosis in Tropical and Subtropical Regions<br>*Mohammad-Ali Assarehzadegan* | **193** |
| **Chapter 7** | The Geographic Mosaic Theory of Coevolution applied to the Neotropical Mutualism *Acacia-Pseudomyrmex*<br>*Sandra Luz Gómez Acevedo* | **213** |
| **Chapter 8** | Potential of *Acacia Melanoxylon* for Pulping<br>*Helena Pereira, Rogério Simões, António Santos, Jorge Gominho, Ana Lourenço and Ofélia Anjos* | **235** |
| **Index** | | **255** |

# PREFACE

*Acacia: Characteristics, Distribution and Uses* opens with a chapter describing the wood of *A. melanoxylon* grown in Portugal in view of determining its technological quality for use in the construction and furniture industry. The characterization includes stem features, wood anatomical characteristics, chemical composition, wood density and mechanical properties.

Next, the authors aim to describe and analyze common characteristics among *Acacia* s.l. species and to trace some parallelisms of their performance throughout several ecosystems that hold such species. It is well-known that *Acacia* s.l. species have the ability to fix atmospheric nitrogen and may modify soil chemistry and physics by enabling microorganisms/soil fauna to alter the microhabitat beneath the tree, and such characteristics are significant in the recovery of ecosystems.

The potential of *Acacia mangium*, an exotic species, for restoration of a degraded land in Mt. Makiling Forest Reserve (MMFR), Philippines was also examined. Results suggested a general trend of changes in *A. mangium* plantation which was once a grassland dominated by *Imperata cylindrica* and *Saccharum officinarum*. Both (stems ha-1) and basal area (m2 ha-1) increased significantly in 2010-2018 (P=0.001).

The authors discuss the way in which NMR spectroscopy applied to the study of gum exudates has become important since the 1990's in Venezuela.

Analytical and structural studies of 23 species belonging to different genera and families have been reported through the combination of classic methodology for carbohydrates and NMR spectroscopy.

The footprints left by evolution in the distribution of characters among current organisms have been one of the main tools in the study of organic evolution. The authors propose that the reconstruction of ancestral character states offers the possibility of knowing the changes suffered by characters in a species throughout evolutionary time.

Pollinosis, also known as pollen allergy, hay fever or seasonal allergic rhinitis is one the most common respiratory disorders throughout the world. The inhalation of *Acacia* pollen is one of the main causes of respiratory allergic diseases in semiarid countries such as Iran, Saudi Arabia, and the United Arab Emirates. This book suggests that the recognition of allergenic components of pollens is essential for component-resolved diagnosis, the design of patient-specific immunotherapy, and the explanation of sensitization mechanisms to various allergens.

The authors analyze *Acacia-Pseudomyrmex* mutualism which includes 15 species of *Acacias* and a group of 10 species of mutualistic ants whose geographical distribution is similar. This relationship is frequently cited as an example of coevolution, a term that has been used to refer to the reciprocal change of interacting species where each of them acts as an agent of natural selection with respect to the other and where the reciprocal selection would result in congruent phylogenies.

The concluding chapter characterizes the *A. melanoxylon* wood pulping performance regarding yield and kappa number as well as the pulp and paper properties. The application of fast spectroscopic technologies for pulp quality determination is also described.

Chapter 1 - *Acacia melanoxylon* R. Br., known as blackwood, is a valued timber species for solid wood products applied in building and carpentry, prized for its heartwood characteristics and pleasant wood aesthetics. *A. melanoxylon* grows naturally in Australia over a wide latitudinal range from north Queensland to Southern Tasmania. The species was introduced in several countries where it showed good growth and adaptation. In Europe it was planted as an ornamental and now spreads in many countries. In

Portugal, blackwood and other *Acacia* spp. were introduced through state afforestation programs in dry and poor sandy soils, and subsequently spread to other regions; it is now present in pure and mixed stands, mainly with maritime pine, with an estimated area about 5000 ha.

The blackwood stem is straight and shows a small taper. The heartwood occupies a large proportion of the stem volume (e.g., over 60% in the lower stem region) while sapwood width is constant with little variation within and between trees. The wood shows distinct growth rings with dense latewood bands. The wood anatomy is characterized by small solitary or radial grouped vessels, homogeneous rays and low parenchyma proportion. The chemical composition shows a low lignin content and a high content of polar extractives in the heartwood. Blackwood is a medium-density hardwood with basic density between 465 kg/m$^3$ and 654 kg/m$^3$. The mechanical properties show potential to supply the industry with valuable hardwood timber.

This chapter describes the wood of *A. melanoxylon* grown in Portugal in view of determining its technological quality for use in the construction and furniture industry. The characterization includes stem features (growth ring, sapwood and heartwood): wood anatomical characteristics, chemical composition, wood density and mechanical properties. A brief summary of the information available on bark is also made regarding chemical features, taking into account its valorisation integrated in a full resource use approach.

Chapter 2 - In 2005, the *Acacia* genus was proposed to be split into several genera, triggering a heated discussion between the botanists of Australia, Africa, and America. This re-typification was ratified in 2006 at the Vienna's Nomenclature Session of the 17th International Botanical Congress (IBC), and later in 2011 in the Melbourne's Nomenclature Section of the 18th IBC. The decisions taken in Melbourne confirm the Australian *Acacias* retain the name, while new names are needed for the African and American species of the Acacieae tribe. Currently, the tribe is formed by the genera *Acacia* (for Australian species), *Vachellia*, *Senegalia*, *Parasenegalia*, *Acaciella*, *Mariosousa*, and *Racosperma*. Despite this division, the species of the tribe along several ecosystems around the globe perform similar characteristics and behaviors. Most species seem to have co-evolved with frequent forest fire events and developed the ability to resprout or activate germination after a fire event. Some *Acacia sensu lato* (*s.l.*) species have also co-evolved with wildlife and seem to adopt resistance to defoliation of seedling browsing. It is well-known that *Acacia s.l.* species have the ability to fix atmospheric nitrogen and, coupled with important leaf litter inputs to the soil, they may modify soil chemistry and physics by enabling microorganisms and soil fauna to alter the microhabitat beneath the tree. Such characteristics are significant in the recovery of ecosystems within ecological successions after disturbance, but may also alter original ecosystems when *Acacias* invade habitats of other native species. These characteristics and others have been found in several species of the tribe Acacieae worldwide. The objective of the chapter is to describe and analyze common characteristics among *Acacia s.l.* species and to trace some parallelisms of their performance throughout several ecosystems that hold such species.

Chapter 3 - There have been a number of debates on the use of exotic species plantations especially for restoration in the Philippines, mainly due to some ecological concerns (e.g., loss of genetic variation). Hence, the potential of *Acacia mangium,* an exotic species, for restoration of a degraded land in Mt. Makiling Forest Reserve (MMFR), Philippines was examined. The compositional and structural changes of the grassland under *A. mangium* plantation over a 20-year and 28-year old study periods were analyzed. Results suggested a general trend of changes in *A. mangium* plantation which was once a grassland dominated by *Imperata cylindrica* and *Saccharum officinarum*. Both (stems ha$^{-1}$) and basal area (m$^2$ ha$^{-1}$) increased significantly in 2010-2018 ($P$=0.001). Increase in the dominance of native overstory trees coupled with a decrease in the dominance of exotic species was also observed. Lastly, results revealed an increase in species richness and species diversity with the presence of more native trees naturally occurring in MMFR (e.g., *Pterocymbium tinctorium and Celtis luzonica)*. These are all attributed to ecological processes including improved soil fertility and microclimate, disturbance, soil seed banks, seed flow of adjacent forests, seed dispersal, and mortality rate along with the N-fixing ability of *Acacia mangium.* This study, therefore, provides insights that *A.*

*mangium* plantation is potential for restoration of degraded lands in the Philippines. However, this may require in-depth ecological knowledge, silvicultural management practices, and continuous field-based monitoring studies to enhance understanding of the management aspect of a restored landscape.

Chapter 4 - NMR spectroscopy applied to the study of gum exudates has become important since the 1990's in Venezuela. Analytical and structural studies of 23 .species, belonging to different genera and families have been reported, through the combination of classic methodology for carbohydrates and NMR spectroscopy. Up to the present, four types of hetero-polysaccharides constituents of these gums have been described: A,B,C and C. Gums of the genera *Acacia, Albizia, Pithecellobium* and *Spondias*, among others, involve heteropolysaccharides of the first type. The core is represented by a ß (1 ---> 3) galactan, in some cases, with residual uronic acids and ß-L-arabinopyranose. These polysaccharides differ basically in the composition and sequences of the L-arabinose side chains. It was evidenced, by 2D-NMR correlation spectroscopy, a carbohydrate-protein linkage in the core of *Acacia tortuosa* gum. On the other hand, characteristic features of type B heteropolysaccharides were recorded, with some variations, for *Sterculia apetala, Cedrela odorata* and *Pereskia guamacho* gum exudates while *Cercidium praecox* heteropolysaccharide corresponds to a C structure, with a ß (1--->4) xylan core. These NMR techniques contributed to identify and confirm relevant structural features of the different gum exudates of species located in Venezuela.

Chapter 5 - The footprints left by evolution in the distribution of characters among current organisms have been one of the main tools in the study of organic evolution. The reconstruction of ancestral character states offers the possibility of knowing the changes suffered by characters in a species throughout evolutionary time; of understanding the origin of the adaptations to which they give rise, and comprehending their function as well as recognizing the way in which the symbiotic interactions that they sustain with other species influence such attributes. The relationships between plants and ants provide numerous examples of mutualism, one of them is the obligate and highly specialized interaction between ants and

*Acacias*. Neotropical ant-*Acacias* are characterized by the presence of extrafloral nectaries, domatia and beltian bodies; traits directly related to their mutualistic association with ants. These traits when analyzed under the perspective of the reconstruction of ancestral character states show that during the transition from the external group to the myrmecophile group the amount of extrafloral nectaries per petiole increased and their shape changed from circular to columnar and canoe-shaped. The stipules grew in length resulting in domatia and cyanogenic glycosides against herbivory were lost. Furthermore, in relation to other morphological characteristics, the leaves were lengthened in a first step and then became smaller. The number of pinnae pairs and pinnules per leaf decreased, while both the length and width of the pinnules increased, as well as the petiole maximum length. Thus, the selection pressure that symbiotic relationships exert on interacting species is reflected in a systematic set of their attributes, which, in the authors' case, were developed in an environmental scenario of warm-humid conditions prevalent in the Neotropics during the Late Miocene and Early Pliocene.

Chapter 6 - Pollinosis, also known as pollen allergy, hay fever or seasonal allergic rhinitis is one the most common respiratory disorders throughout the world. The main cause of pollinosis is pollen of different species of trees. Different species of *Acacia* are common throughout tropical and subtropical regions of Asia, Africa, Australia, and America with hot and humid climates, where it is planted as a shade and/or ornamental tree or for binding sand. Generally, pollens from different members of the Fabaceae family, particularly *Acacia* Spp. and Prosopis Spp., have been reported as an important source of pollinosis in the United States, European countries, and Asia. Moreover, the inhalation of *Acacia* pollen is one of the main causes of respiratory allergic diseases in semiarid countries such as Iran, Saudi Arabia, and the United Arab Emirates, where the frequency of sensitization ranges from 25% to 48%. The crude extracts from trees pollen are usually composed of several protein and glycoprotein components. The protein molecules are rapidly diffused after contact with mucus membranes of humans. The recognition of allergenic components of pollens is essential for component-resolved diagnosis, the design of patient-specific immunotherapy, and the explanation of sensitization mechanisms to various allergens. Immunochemical analysis of the protein profile of *Acacia farnesiana* (*Vachellia farnesiana*) pollen indicated several components ranging from 12 to 105 kilodalton (kDa). Moreover, the 15, 45, 50, 66 and 85 kDa proteins are recognized as dominant IgE-binding components of *Acacia* pollen. Despite a high prevalence of sensitization to *Acacia* pollen in semiarid countries, there is little information about the molecular characterization of *A. farnesiana* pollen allergens. In one recent report, a new allergen from *Acacia* pollen was introduced for the first time. This allergen was named Aca f 1 in accordance with the International Union of Immunological Societies (IUIS) Allergen Nomenclature Subcommittee. Aca f 1 is a member of the Ole e 1-like protein family. Aca f 2, other new allergen from *A. farnesiana* pollen, is an allergenic member of the profilin family. Profilins are 12- to 15-kDa monomeric proteins belonging to a ubiquitous family of actin-binding proteins. They are prominent allergens in the pollens of trees, grasses, and weeds, and many fruits and vegetables. Cross-reactivity among allergen sources occurs when the allergenic proteins from one source of allergen are similar to the allergenic proteins in another source of allergen. Cross-reactivity among *Acacia* pollen components with *Lolium perenne* pollen allergens has been described. Moreover the results of earlier studies indicated a significant IgE cross-reactivity between *Acacia* and *P. juliflora* (Mesquite) pollens. In general, the knowledge of pollen cross-reactivity is crucial to diagnostics as well as formulation of immunotherapy vaccines. Furthermore a high degree of sequence similarity (90%) was detected between Aca f 1 and Sal k 5 (Allergenic profilin of *Salsola kali* pollen). Besides the results of amino acid sequence similarity analysis revealed that Aca f 2 has a high degree of identity with the selected profilins from the most common allergenic regional plants, particularly Pro j 2 (*P. juliflora* profilin) (96%).

Chapter 7 - The symbiotic associations between plants and ants are one of the classic examples of mutualism in nature. This kind of relationship achieves a high degree of sophistication in the myrmecophytic plants, which are characterized by offering permanent shelter and food for their resident ants. In return, the ants protect the plants for the herbivory and confer additional benefits such as protection against pathogenic fungi; the removal of lianas; the addition of nutrients (principally nitrogen) through waste decomposition; and the absorption of $CO_2$. The relationship between swollen thorn *Acacias* and their associated ants has been one of the best known symbiosis so far. The Neotropical *Acacia-Pseudomyrmex* mutualism includes 15 species of *Acacias* and a group of 10 species of mutualistic ants whose geographical distribution is similar. This relationship is frequently cited as an example of coevolution, a term that has been used to refer to the reciprocal change of interacting species, where each of them acts as an agent of natural selection with respect to the other, and where the reciprocal selection would result in congruent phylogenies. However, several reports indicate that the relationship between these taxa does not correspond to the original proposal of strict coevolution, since throughout its geographical distribution, a single ant-*Acacia* species can be inhabited by two or more species of mutualistic ants and even non-mutualistic ants, and have referred to this system as an example of diffuse coevolution. Nevertheless, this kind of relationship shows that although at a population level, relations imply a certain specialization towards one or another species of ant, at a global level, when considering the entire geographical distribution, the evolutionary unit of interaction can include more than one pair of species. This generates a whole range of interrelations, pointing out that the coevolutionary process between these taxa is highly dynamic and corresponds to the proposal of the geographical mosaic theory.

Chapter 8 - Most of the fibre raw materials used by the pulp and paper industry are from a small number of tree species. For instance, *Eucalyptus* and *Pinus* species are the major industrial pulpwood sources obtained from forests that are characterized by a relatively low biodiversity. The large monoculture areas also increase environmental risks such as those related to biotic attacks or forest fires. Diversification of industrial fibre sources has therefore been a matter of research and the characterization of different raw materials has been made in view of their pulping potential. *Acacia melanoxylon* R. Br. (blackwood) grows well in Portugal in pure or mixed stands with *Pinus pinaster* Aiton, and is valued as a timber species with potential for sawmills. In addition, the wood anatomical and chemical characteristics also allow to consider the species as an alternative raw material for the pulp industry. *Acacia* species, with their relatively short, flexible and collapsible fibres, have potential to produce papers with good trade-offs between light scattering/tensile strength and smoothness/tensile strength, at low energy consumption in refining. The pulping and paper making potential of blackwood has been studied by several authors who showed an overall good pulping aptitude under the same experimental conditions of kraft pulping as used for eucalypt pulping, with pulp yields ranging between 47% and 58%. The presence of heartwood should be taken into account because it decreases the raw-material pulping quality due to the higher extractives content. Heartwood proportion should therefore be considered as a quality variable when using *A. melanoxylon* wood in pulp industries. This chapter characterizes the *A. melanoxylon* wood pulping performance, regarding yield and kappa number, and the pulp and paper properties. The application of fast spectroscopic technologies for pulp quality determination is also described.

In: *Acacia*
Editor: Aide Matheson

ISBN: 978-1-53614-237-2
© 2018 Nova Science Publishers, Inc.

*Chapter 1*

# STEM AND WOOD CHARACTERIZATION OF *ACACIA MELANOXYLON* AS AN INTRODUCED SPECIES IN EUROPE

*Ofélia Anjos*[1,2,3,*], *António J. A. Santos*[2], *Isabel Miranda*[2], *Sofia Knapic*[2,4,5] *and Helena Pereira*[2]

[1]Instituto Politécnico de Castelo Branco, Castelo Branco, Portugal
[2]Centro de Estudos Florestais, Instituto Superior de Agronomia, Universidade de Lisboa, Lisboa, Portugal
[3]Centro de Biotecnologia de Plantas da Beira Interior, Castelo Branco, Portugal
[4]SerQ-Centro de Inovação e Competências da Floresta, Sertã, Portugal
[5]Institute for Sustainability and Innovation in Structural Engineering, Faculdade de Ciências e Tecnologia, Universidade de Coimbra, Portugal

## ABSTRACT

*Acacia melanoxylon* R. Br., known as blackwood, is a valued timber species for solid wood products applied in building and carpentry, prized

[*] Corresponding Author Email: ofelia@ipcb.pt

for its heartwood characteristics and pleasant wood aesthetics. *A. melanoxylon* grows naturally in Australia over a wide latitudinal range from north Queensland to Southern Tasmania. The species was introduced in several countries where it showed good growth and adaptation. In Europe it was planted as an ornamental and now spreads in many countries. In Portugal, blackwood and other *Acacia* spp. were introduced through state afforestation programs in dry and poor sandy soils, and subsequently spread to other regions; it is now present in pure and mixed stands, mainly with maritime pine, with an estimated area about 5000 ha.

The blackwood stem is straight and shows a small taper. The heartwood occupies a large proportion of the stem volume (e.g., over 60% in the lower stem region) while sapwood width is constant with little variation within and between trees. The wood shows distinct growth rings with dense latewood bands. The wood anatomy is characterized by small solitary or radial grouped vessels, homogeneous rays and low parenchyma proportion. The chemical composition shows a low lignin content and a high content of polar extractives in the heartwood. Blackwood is a medium-density hardwood with basic density between 465 kg/m$^3$ and 654 kg/m$^3$. The mechanical properties show potential to supply the industry with valuable hardwood timber.

This chapter describes the wood of *A. melanoxylon* grown in Portugal in view of determining its technological quality for use in the construction and furniture industry. The characterization includes stem features (growth ring, sapwood and heartwood): wood anatomical characteristics, chemical composition, wood density and mechanical properties. A brief summary of the information available on bark is also made regarding chemical features, taking into account its valorisation integrated in a full resource use approach.

# INTRODUCTION

*Acacia melanoxylon* R. Br., usually named blackwood or Australian blackwood is a species of the genus *Acacia*, belonging to the subfamily Mimosoideae of the family Fabaceae (Seigler, 2003; Miller and Murphy, 2005). *A. melanoxylon* has a native distribution in Australia through the tablelands and coastal escarpments of southeast Queensland, New South Wales and Victoria to the Mount Lofty Ranges in South Australia and to southern Tasmania (Doran and Gunn, 1987). According to Le Floc'h (1991), *A. melanoxylon* was introduced to several countries in the western Mediterranean basin in the 1800s and has subsequently naturalized in

Algeria, France and Portugal. Now *A. melanoxylon* exists in Europe (Portugal, Spain, France, Italy, United Kingdom), South Africa, India, USA (California, Hawaii), South America (Argentina, Chile) and in some regions of New Zealand (Fernandes et al., 2008; Marchante et al., 2003).

*A. melanoxylon* grows fast and is a long-lived species; in Tasmania some trees have been estimated to be more than 200 years old. *A. melanoxylon* is a nitrogen fixing species, able to grow on a variety of soils, prolifically produces seeds with extreme longevity and high germinability from an early age and is also able to spread from root suckers (Jennings, 1998). *A. melanoxylon* varies from a small shrub to a large tree with 10-20 m height and 0.5 m in diameter. In an open stand the small and medium-sized trees could branch to ground levels (https://www.cabi.org/isc/datasheet/2329).

In Portugal, *A. melanoxylon* trees grow spontaneously and are distributed over the country, especially in the Atlantic zone (Leite et al., 1999). It is now classified as an invasive species (Decree-Law nr. 565/99, of 21 December). In fact, these species propagate vegetatively and produce many seeds with a high dispersal efficiency that remain viable in the soil for more than 50 years and germinate well after space opening or fire occurrences (Arán et al., 2013; Richardson and Kluge, 2008).

In Europe there are no commercial plantations. In Portugal, *A. melanoxylon* is well adapted to soil and climate and can grow in pure or mixed stands with other species namely *Pinus pinaster* Aiton with a good annual increment (0.89 cm/year) (Rucha et al., 2011), when compared with maritime pine (0.58 to 0.85 cm/year) (Tavares et al., 2004) and eucalyptus (0.84 to 0.96 cm/year) (Tomé et al., 2001).

Blackwood is prized for cabinet work, panelling, inlays, bent work and staves. Availability of large logs is limited and today the timber is mainly used for sliced veneer, especially on particleboard for cabinet work and furniture, woodcraft and it is universally regarded as an excellent interior feature timber. The wood has good acoustic qualities and is suitable for violin backs. Small diameter, fast-grown logs do not develop the growth stresses of some eucalypt species and good sawn timber conversion can be expected from trees grown on 40-50-year rotations (Whibley and Symon, 1992). Blackwood has also good pulpwood potential giving acceptable pulp yields and paper properties favourable for fine papers, as discussed in the next chapter.

## STEM QUALITY

The stem quality of timber trees is given by parameters such as straightness, taper and pith eccentricity that are directly related to sawmilling aptitude and product yield, as well as by the within-tree knot architecture and growth ring homogeneity that are determining for establishing the wood component value. Another important stem quality parameter is the within tree development of heartwood and sapwood that are linked to the wood valorisation for some applications.

Heartwood is different from sapwood regarding chemical composition, density and some physical and technological properties. Heartwood is desirable for wood components since in general it has higher durability, better dimensional stability and aesthetics. Therefore, the heartwood content and its within-tree development are important when the tree is directed for timber applications, e.g., the yield of heartwood sawn products is directly related to the heartwood diameter within the log (Pinto et al., 2004). This is the case for *A. melanoxylon* that has a wood with high aesthetical value similar to walnut, mahogany and teak (Bradbury et al., 2011; Nicholas and Brown 2002).

The transformation of sapwood into heartwood is characterized by the death of parenchyma cells, development of tyloses in the vessels and biosynthesis of nonstructural compounds, leading to an important accumulation of extractives (Bamber, 1976; Hillis, 1987). Heartwood development varies between and within species and has been related to growth rates, stand and tree biometry, site conditions and genetics (Bamber and Fukazawa 1985; Hillis 1987; Taylor et al., 2002). Heartwood is impregnated with extractives namely polyphenols, enabling reduction of shrinkage and swelling, increasing durability and usually conferring a darker colour (Hillis 1987; Taylor et al., 2002). They do not necessarily give a higher mechanical resistance, but they can prevent warping in the central part of juvenile wood. Heartwood usually has lower permeability than sapwood, which suggests that it is more difficult to dry and less penetrable by preservatives, paint and dyes (Zhang, 1997).

Few works have characterized the heartwood and sapwood development in *A. melanoxylon,* even if the wood value of the trees is given primarily by their high heartwood content and the best log prices are usually paid for the dark red-brown colored heartwoods with a good diameter and a round shape (Harrison, 1974, Harrison, 1975, Igartúa et al., 2017, Knapic et al., 2006, Zwaan, 1982). The most extensive study on *A. melanoxylon* heartwood development was carried out by Knapic et al., (2006) on mature trees growing in four stands in Portugal, with data collected at different height levels along the stem from base to top.

In the lower part of the tree, heartwood represented 67% of the stem cross-sectional area and totalled 66% of the wood volume while the sapwood has an average radial width of 21% that remains rather constant along the stem i.e., the heartwood profile follows the stem profile (Knapic et al., 2006). Figure 1 displays the variation of the stem cross-sections with a clear distinction of heartwood and sapwood for the base, 5, 15, 35, 50, 65 and 90% height levels. Similar values were obtained by Harrison (1974, 1975) for blackwood trees in different locations in South Africa.

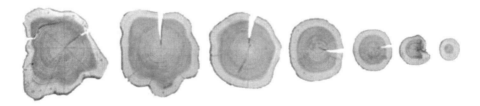

Figure 1. Variation of the stem and heartwood cross-sections for the base, 5, 15, 35, 50, 65 and 90% height levels (from left to right). Source: Knapic et al., 2006.

These results showed the high quality of the *A. melanoxylon* trees regarding heartwood content and therefore their suitability for sawn timber for all types of valued applications (Knapic et al., 2006; Maslin and Pedley, 1982a; Maslin and Pedley, 1982b).

As regards annual growth, *A. melanoxylon* is considered to be a fast growing species. Tavares et al., (2014) reported values of mean annual ring width of 6 mm in Portugal. Medhurst et al., (2003) reported diameter growth rates of up to 1.4 cm/year in 7-year-old trees, though Jennings et al., (2000) in 7–21-year-old natural *A. melanoxylon* swamp forests referred lower average growth diameter of 0.6–0.7 cm/year.

Annual growth ring width for African *Acacia* species was reported as ranging from 2 to 20 mm with a mean of 7.5 mm (Gourlay and Kanowski 1991) and for *Acacia karroo* average annual growth ring width varied between 6 and 7 mm in stands and between 3.5 to 15.2 mm for individual trees (Gourlay and Barnes 1994).

Age is responsible for around 49% of latewood volume variation, and the variation between sites is also relevant and explains around 15% of the total variation (Rucha et al., 2011).

The pith eccentricity of blackwood growing in Portugal was studied by Santos et al., (2013). The pith eccentricity values were generally very low, and the higher values were found at the top and at the base level.

The vertical heartwood shape accompanied the stem shape and up to 50% of tree height, stem taper and heartwood taper were similar at an average 9 mm/m while for the upper part of the tree, the stem and heartwood tapers increased, especially after the 65% height level, with a significantly higher conicity for the heartwood (Knapic et al., 2006).

## WOOD ANATOMY

Macroscopically the growth rings in *A. melanoxylon* wood are generally indistinct or not well defined in the majority of the wood samples, and their differentiation is therefore only possible at a microscopic level (Quirk 1983; Wilkins and Papassotiriou 1989; Santos et al., 2018). At this level, and as shown in Figure 2 (A, B and C), *A. melanoxylon* wood is characterized by distinct growth rings marked by a latewood band composed of dense fibres with smaller radial dimension and thicker cell wall in comparison to the earlywood fibres (Santos et al., 2018). In *A. melanoxylon* trees from Asia and Australia, the rings are demarcated by 10-15 rows of flattened fibres, with a thickened wall (Quirk 1983). The latewood usually has smaller and fewer vessels compared to the earlywood (Figure 2 C) and sometimes it shows absence of vessels (Wilkins and Papassotiriou, 1989; Quirk 1983). Hillis et al., (1987) observed dark lines within the heartwood that were not related to annual rings but associated to the abnormal formation of tension wood linked with changes in coloration.

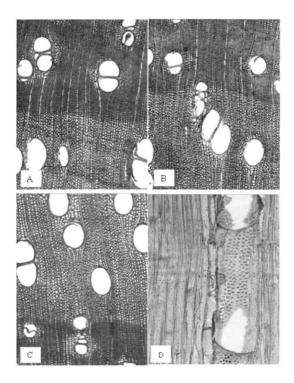

Figure 2. Transverse section (A, B and C) (50 x) and radial section (D) (200x) of a wood sample of *A. melanoxylon*. Photos from Ofélia Anjos, 2018.

The wood shows a diffuse porosity with isolated vessels or vessels grouped in two (Figure 2, A, B, C and D). The vessel percentage (area %) ranged between 6% and 15%, with a low porosity near the pith (Santos et al., 2018). The vessels are small mostly less than 200 µm in diameter with simple perforation plates. Quirk (1983) found values around 69.2 µm of vessel diameter and a low density of 5/mm². Monteoliva and Iguartúa (2010) reported small to medium sized pores with an average diameter of 69.3 μm (±22.3), mostly solitary, although with radial series of 2 to 4 vessels and grouped. Wilkins and Papassotiriou (1989) reported larger vessel diameters but similar frequencies (on average 5/mm²) and related the variation of vessel diameter and frequency to differences between sites. Similar conclusions were reported by Rodrigues et al., (2007) and Santos et al., (2018).

Parenchyma is observed in irregular bands, often indistinct, or in marginal discontinuous bands (Quirk 1983). The paratracheal parenchyma is often composed by few cells in contact with the vessels, vasicentric (Figure 2 A, B and D), confluent, and broad bands in the most external positions (Santos et al., 2018; Monteoliva and Iguartúa 2010). Monteoliva and Iguartúa (2010) observed also axial parenchyma in long series and the presence of scarce series of cubic crystals.

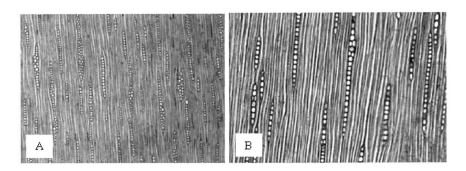

Figure 3. Tangential section of a wood sample of *A. melanoxylon* (A:50 x: B:100x) Photo from Ofélia Anjos, 2018.

Rays are not storied, homocellular (Figure 3), with 3 cells or less wide and without crystals (Quirk 1983). Usually uniseriate rays are in majority but with some bi- and tri-seriate rays (Santos et al., 2018; Monteoliva and Iguartúa, 2010). However, Quirk (1983) reported 80% multiseriate and only 20% uniseriate in *A. melanoxylon* from Australia.

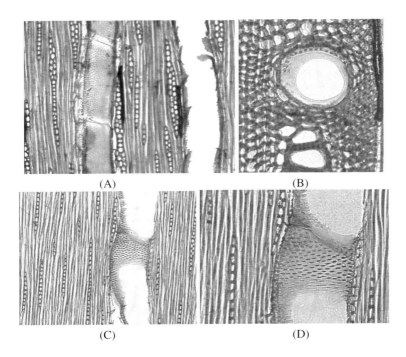

Figure 4. Vessels of a wood sample of *A. melanoxylon*: A-Tangential section, 100x; B – Transversal section with vessel overture, 200x; C - Tangential section with vessel ponctuations, 100x; D - Tangential section with vessel ponctuations, 200x. Photos from Ofélia Anjos, 2018.

Monteoliva and Iguartúa (2010) reported the presence of vasicentric tracheids. Vessel with an oblique overture and with alternating areolated ponctuations (Figure 4).

For *A. melanoxylon* growing in Portugal, wood fibre length and wall thickness varied between 0.90-0.96 mm and 3.45-3.89 µm respectively (Tavares et al., 2011; Anjos et al., 2011; Santos et al., 2018). The within tree variation of fibre biometry showed an axial decrease in fibre length from bottom to top of the tree, and a radial increase from pith to bark of fibre length and wall thickness (Tavares et al., 2011). Santos et al., (2006) reported fibre length values of 0.650 mm for 22-year-old *A. melanoxylon* trees growing in Portugal. For *A. melanoxylon* from Buenos Aires, short fibres of 655.6 µm in length were measured, with diameter around 12.9 µm and thin walls of 2.2 µm thickness with very few simple pits (Monteoliva and Iguartúa, 2010). For trees in Chile, Campos et al., (1990) reported higher values of fibre length of 1.45 mm.

Some anatomical characteristics of *A. melanoxylon* wood vary with the region (Rodrigues et al., 2007). Wilkins and Papassotiriou (1989) reported that a number of anatomical characteristics of *A. melanoxylon* wood were positively related to latitude in eastern Australia, namely vessel member length, proportion of fibres and proportion of multiseriate rays and some were negatively related to latitude namely vessel frequency, vessel diameter, abundance of crystals and the proportion of uniseriate rays, vessels and axial parenchyma.

## WOOD CHEMISTRY

The wood chemical composition of *A. melanoxylon* was reported as 0.35-0.51% ash, 3.2-7.8% extractives, 17.5-22.4% Klason lignin and 63.8-72.5% polysaccharides (Lourenço et al., 2008; Santos et al., 2006; Santos et al., 2012; Mansilla et al., 1991). Most of the extractives correspond to ethanol and water soluble compounds (representing 86% of the total extractives and 2.7% of the wood) while the lipophilic extractives (soluble in dichloromethane) represent a minor proportion (14% of the extractives and 0.4% of the wood) (Santos et al., 2006) (Table 1).

Heartwood and sapwood were characterized separately since the stem of *A. melanoxylon* mature trees contains a considerable heartwood proportion (Knapic et al., 2006). Heartwood has more extractives than sapwood (7.4%-9.5% and 4.0%–4.2%, respectively) and the heartwood-to-sapwood extractives ratio varied between trees from 1.9 to 2.3 (Lourenço et al., 2008). The higher proportion of extractives in the heartwood is mainly due to an accumulation of ethanol-soluble compounds, i.e., they represent more than 70% of the total extractives (40% in sapwood). The content of dichloromethane-soluble extractives is negligible (0.4% of wood), while ethanol-soluble and water-soluble extractives correspond on average to 5.9% and 1.8% in heartwood, and 2.0% and 1.7% in sapwood, respectively (Lourenço et al., 2008).

## Table 1. Chemical composition of the wood from *Acacia melanoxylon* (% of dry mass)

| Analysis | Mansilla et al., (1991)[a] | Santos et al., (2006) | Lourenço et al., (2008)[c] Sapwood | Lourenço et al., (2008)[c] Heartwood | Santos et al., (2012) |
|---|---|---|---|---|---|
| Ash | 0.36 | | | | 0.35-0.51 |
| Total extractives | 4.1 | 3.2 (0.41) | 4.0-4.2 | 7.4-9.5 | 5.29-7.59 |
| DCM | | 0.43 (0.13) | 0.3-0.4 | 0.3-0.5 | |
| EtOH | | 1.81 (0.32) | 1.9-2.2 | 5.3-7.3 | |
| H$_2$O | | 0.93 (0.08) | 1.7-1.8 | 1.7-2.0 | |
| Total lignin | 26.4 | 17.5 (1.59) | 20.9-22.2 | 20.5-20.9 | 24.58-28-23 |
| Klason lignin | | | 18.1-19.8 | 17.9-18.7 | 19.46-22.32 |
| Soluble lignin | | | 2.4-2.8 | 2.2-2.7 | |
| Cellulose | 39.6 | | | | |
| Holocellulose | 72.5 | | | | |
| Sugar composition | | | | | |
| Arabinose | | | 1.1-1.8 | 0.8-1.6 | |
| Xylose | | | 15.6-18.7 | 15.1-18.91 | 16.95-18.79 |
| Mannose | | | 3.3-3.7 | 2.8-3.9 | |
| Galactose | | | 0.7-0.8 | 0.6-0.8 | |
| Glucose | | | 46.1-50.8 | 46.6-50.2 | 45.52-49.59 |
| Acetic acid | | | | | 1.11-1.97 |

a – mean value; b – mean value from wood chips at the bottom and at the top of the tree; c – average 4 sites in Portugal - 20 trees; d – average 4 sites in Portugal – 12 trees.

There are few published references on the wood chemical composition of other *Acacia* species: 0.6 ash, 5.8% total extractives, 19.4-21.7% Klason lignin, and 69.8-76.1% polysaccharides for *A. auriculiformis* (Collins et al., 1990; Jahan et al., 2008) as well as 5.3-8.3% extractives and 22.4-42.6% Klason lignin also for *A. auriculiformis* (Muhammad et al., 2018); 3.5% extractives and 18.2% Klason lignin for *A. dealbata* (Santos et al., 2006); and 0.5 ash, 4.5% ethanol/toluene extractives, 25.6- 27.6% Klason lignin and 61.0-71.5% polysaccharides for *A. mangium* (Neto et al., 2004; Pinto et al., 2005a; Jahan et al., 2008; Çetinkol et al., 2012).

The monomeric composition of *A. melanoxylon* wood polysaccharides shows that glucose is the main monosaccharide unit, ranging from 67.2% to 71.1% of the total units; xylose represents 22.5% to 28.5% of the total units and together arabinose and galactose represent 2.6% of the total content of sugars; the degree of acetylation is very high corresponding to 1.7-2.9% of total monomers (Lourenço et al., 2008; Santos et al., 2012). The wood hemicelluloses of *A. melanoxylon* are therefore constituted mostly by acetylated xylans. There are no difference between heartwood and sapwood regarding the carbohydrate composition. Similar results were reported for the sapwood and heartwood of *A. crassicarpa* and *A. mangium* with cellulose as the predominant polysaccharide, and xylans as the main non-cellulosic polysaccharides with glucuronic and galacturonic acidic sugar units representing 15-23% of the total amount of non-cellulosic units (Willför et al., 2005).

The lignin of *Acacia* species is representative of hardwood lignins that are mainly constituted by guaiacyl (G) and syringyl (S) units with a small proportion of hydroxyphenol (H) units (Lourenço and Pereira, 2018). Lignin from *A. mangium* presented a H:G:S relation of 1:16:16 (by NMR) and 1:21:12 (by permanganate oxidation), with a S/G of 0.98 and 0.56, showing a high degree of condensation and low content of β-O-4′ inter monomeric linkages (Pinto et al., 2005b). Syringyl ratios (S/S+G) ranged from 0.4 to 0.7 in *A. mearnsii*, *A. mangium*, *A. auriculiformis* and hybrids (Nawawi et al., 2017; Huang et al., 2016) and S/G ratios were 1.33 for *A. mangium* heartwood and 0.79 for *A. mangium* sapwood (Çetinkol et al., 2012).

The lipophilic fraction of wood extractives, even present in only small amounts was characterized: sterols, fatty acids, long chain aliphatic alcohols and aromatic compounds are the main families of compounds (Freire et al. 2005, 2006). Fatty acids represent the major lipophilic components: for *A. mangium* wood, free fatty acids varied between 690-1706 mg/kg o.d. wood, composed mostly of C24, C26 and C28 acids with a smaller proportion of C16-C18 acids (Freire et al., 2006; Pietarinen et al., 2004); for *A. crassicarpa* wood, the total amount of free fatty acids varied between 140 - 330 mg/kg o.d. wood with C24 as the dominant acid and with C16 and C18:2 acids present in large amounts (Pietarinen et al., 2004). In the heartwood and sapwood of an *Acacia* hybrid, the lipophilic extractives contained, respectively 670-950 and 300-660 mg/kg of the dry extract with mainly the C24 and C22 acids (Soon and Chiang, 2012).

Sterols are also found in significant amounts in lipophilic extracts of *Acacia* woods. *A. melanoxylon* wood contains high amounts of free sterols (620 mg/kg of dry wood): two Δ7 sterols, spinasterol (326 mg/kg of dry wood) and dihydrospinasterol (275 mg /kg of dry wood) are the main sterols identified (Freire et al. 2005). Steryl glucosides are also found although in lower amounts: spinasteryl glucide (80 mg/kg of dry wood) and dihydrospinasteryl glucoside (79 mg/kg of dry wood) (Freire et al., 2005). Sterols are also among the major lipophilic components of *A. dealbata*, *A. retinodes*, *A. mangium* and *A. crassicarpa* woods (Freire et al., 2005; 2006; Pietarinen et al., 2004) as well as in an *Acacia* hybrid (heartwood contained 250-300 mg/kg of dry wood and sapwood contained 160-290 mg/kg of dry wood) (Soon and Chiang, 2012).

Two quinones were isolated from *A. melanoxylon* heartwood ethanol extracts, the yellow quinone as 2,6-dimethoxy-p-benzoquinone and the red quinone 2-methyl-6-methoxy-furano-benzoquinone (acamelin) which are responsible for the allergy-inducing properties of this species (Hausen and Schmalle, 1981; Schmalle and Hausen, 1980; Hausen et al., 1990). Large amounts of alkaloid-like substances such as β-phenylethylamine and tryptamine were also isolated from *A. melanoxylon* heartwood (Simes et al., 1959).

The composition of hydrophilic extractives of wood of several *Acacia* species has been studied in phytochemical surveys yielding evidence for the wide distribution of flavonoids in this genus. Two flavan-3,4-diols, melacacidin and isomelacacidin, were isolated in higher amounts in *A melanoxylon* heartwood (Barry et al., 2003; Tindale and Roux, 1974).

The hydrophilic extractives of *A. melanoxylon* aerial parts (wood, bark and leaves) were characterized by Luis et al., (2012) and shown to contain a considerable amount of total phenolics (100-138 mg gallic acid equivalent/g of dry extract), and a low content of flavonoids (22-99 mg quercetin equivalent/g of dry extract) and of alkaloids (4-18 mg pilocarpine nitrate equivalents/g of dry extract). Eight phenolic compounds were identified and quantified in the hydrophilic extract of *A. melanoxylon* wood: hydroxybenzoic acids (3.5-5.8 mg ellagic acid/g of the extract, 1.2 mg vanillic acid/g of the extract), hydroxycinnamic acids (1.3-3.2 mg p-coumaric acid/g of the extract, 2.3-7.6 mg ferulic acid/g of the extract, 2.9 mg syringic acid/g of the extract, 0.9-1.8 mg caffeic acid/g of the extract and 3.3-7.1 mg chlorogenic acid/g of the extract) and flavonoids (0.77-4.36 mg quercentin/g of the extract).

The hydrophilic extract of *A. melanoxylon* wood showed moderately good results to act as a free radical scavenger with $IC_{50}$ values 11.9-13.5 μg/mL as compared to well-known antioxidant standards like gallic acid (2.8 μg/mL), catechin (5.5 μg/mL), butylhydroxytoluene (BHT) (17.0 μg/mL), and ascorbic acid (9.4 μg/mL) (Luis et al., 2012).

## WOOD DENSITY AND MECHANICAL PROPERTIES

Blackwood is a medium density hardwood with basic density ranging between 465 and 671 kg/m$^3$ (Harris and Young, 1988; Clark et al., 1992). Mean density values have been reported by several authors: 564 kg/m$^3$ by Igartúa et al., (2009), 650 kg/m$^3$ by Machado and Cruz (2005) and between 432 and 658 kg/m$^3$ by Santos et al., (2013). Searle and Owen (2005) referred a mean whole-tree basic density of 576 kg/m$^3$ for 8-year-old *A. melanoxylon* trees.

Basic density generally increases with age, although there is a large between-tree variation (Harris and Young, 1988). The basic density varied considerably over the radial profile, increasing with cambial age (Nicholas and Brown, 2002; Igartúa et al., 2009). A large variation of basic density was also reported between sapwood (494–740 kg/m$^3$) and heartwood (583–987 kg/m$^3$) (Aguilera and Zamora, 2009; Igartúa et al., 2009).

The pattern of axial variation of density within the stem has been reported with contradictory results: Santos et al., (2013) reported for trees in Portugal the trend of higher values near the top than near the base, Igartúa et al., (2009) referred a significant decrease with height in Argentinian trees, while Nicholas and Brown (2002) observed very little variation with height in New Zealand blackwood trees.

Blackwood is highly appreciated due to medium bending properties, high crushing strength and resistance to impact, which are all important properties for structural uses (Carvalho, 1977; Nicholas and Brown, 2002). Igartúa et al., (2009) reported the following mean strength values: 89.9 N/mm$^2$ in bending, 49.5 N/mm$^2$ in compression parallel to the grain, 11.7 N/mm$^2$ in compression perpendicular to the grain, 11.6 N/mm$^2$ in shear parallel to the grain, 45.3 N/mm$^2$ hardness and a modulus of elasticity in bending of 10926 N/mm$^2$. Machado and Cruz (2005) reported a mean bending strength of 146 N/mm$^2$, a modulus of elasticity of 14200 N/mm$^2$ and an axial compression strength of 61 N/mm$^2$. Machado et al., (2014) reported 139 N/mm$^2$ for bending strength, 14100 N/mm$^2$ for the modulus of elasticity and 61 N/mm$^2$ for compression strength parallel to grain. The most important factor to explain the variation of mechanical properties was the variability between individual trees. However, within the tree, the radial variation is highly significant (Machado et al., 2014).

The European blackwood is a medium strength wood with mechanical properties similar to those reported for blackwood in Australia and New Zealand: 115–130 N/mm$^2$ for bending strength, 10700 N/mm$^2$ for modulus of elasticity and 60–63 N/mm$^2$ for compression strength parallel to grain (Bolza and Kloot, 1963; Haslett 1986). Blackwood also compares favourably to the prized teak wood that shows 141 N/mm$^2$ for bending strength, 13200–14400 N/mm$^2$ for modulus of elasticity and 50 N/mm$^2$ for compression strength parallel to grain (Miranda et al., 2011).

## BARK CHARACTERIZATION

Literature on bark of *Acacia* species is very scarce and there is only some information on specific chemical features namely focusing on extractives.

For *A. melanoxylon* bark, the non-polar extractives (dichloromethane soluble) account for 2.04% of the bark, which is a value in the range reported for barks of other *Acacia* species (0.92% for *A. longifolia*, 2.0% for *A.*

*dealbata*, 2.71% for *A. retinodes*) and much higher than that found in the corresponding wood (0.5% for *A. melanoxylon* wood) (Freire et al., 2005).

These dichloromethane soluble extracts of *A. melanoxylon* bark contain free sterols, long-chain fatty alcohols and monoglycerides as major components, along with fatty acids and alcohols (Freire et al., 2005, 2007). The following contents were reported: 212.6 mg of free sterols/kg dry bark, spinasterol (95.8 mg/kg dry bark), dihydrospinasterol (102.9 mg/kg dry bark) and steryl glucosides (262 mg/kg dry bark) including spinasteryl glucoside (170.5 mg/kg dry bark) and dihydrospinasteryl glucoside (92.0 mg/kg dry bark) (Freire et al., 2005). Free sterols and Δ7 steryl glucosides were also identified in the lipophilic extracts of *A. dealbata* and *A. retinodes* barks (Freire et al., 2005).

Phenolic components are also present in the *A. melanoxylon* bark dichloromethane extracts. Long-chain n-alkyl ferulates and coumarates, n-alkyl cinnamates, with caffeates representing the major fraction, account for 4.0 g/kg of dry bark, with tetracosanyl (655 mg/kg), hexacosanyl (2764 mg/kg) and octacosanyl (123 mg/kg) caffeates as the most abundant components (Freire et al., 2007). The same phenolic components were also identified in the bark of *A. longifolia*, and *A. retinodes* (Freire et al., 2007).

The bark of several *Acacia* species (e.g., *A. mearnsii*, *A. mangium*, *A. auriculiformis*) has been used as an important source of tannins for producing tanned leather and for the formulation of waterproof adhesives for wood composites. For instance, the bark of *A. mangium* contains 20.0-37.9% of total extractives, of which 13.7-17.7% of the bark are polyphenolic compounds (Makino et al., 2009; Hoong et al., 2010). The bark of *A. auriculiformis* contains about 28.6% of extractives and the polyphenolic content corresponds to 10.9% of the bark (Makino et al., 2009). In the bark of *A. mearnsii* the total phenolics content amounts to 21.7-38.0% of the bark (Duan et al., 2005). These extracts were found to be reasonably reactive towards formaldehyde (as shown by a high Stiasny number) e.g., the Stiasny number of *A. mangium* extract was 70 - 90% (Hoong et al., 2009, 2010). The reactive phenol content in *A. confusa* bark extracts was 46.5-51.6% (Lee and Lan 2006).

The structure of the condensed tannins from *Acacia* species bark consist of combinations of flavan-3-ol units such as profisetinidin (PF), prorobinetinidin (PB) and prodelphinidin (PD) (Figure 5) (Ishida et al., 2005): *A. mangium* condensed tannins consist predominantly of prorobinetinidin combined with profisetinidin and prodelphinidin (Hoong et al., 2010) while the condensed tannins from the bark of *A. confusa* include propelargonidin and procyanidin (Wei et al., 2010). Mimosa tannins (*A. mearnsii*) is predominantly composed of prorobinetinidins (Pasch et al., 2001).

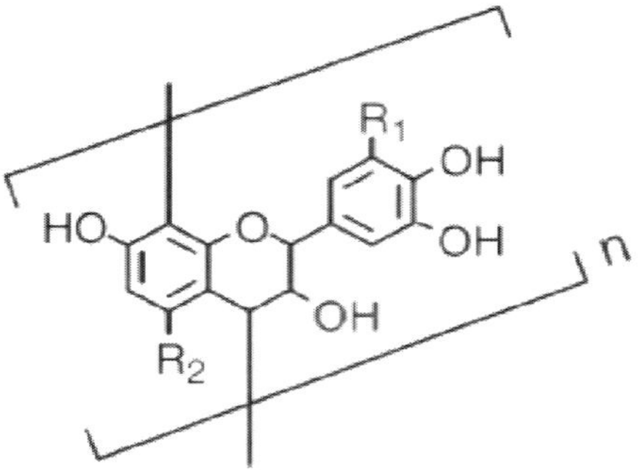

Figure 5. Proposed structures of condensed tannins present in *A. mangium* extracts. R1 = H, profisetinidin (PF), $R^2$ = H, prorobinetinidin (PB); R1, R2 = OH, prodelphinidin (PD) (Ishida et al., 2005).

## APPLICATIONS

Blackwood can be classed in "an elite group of species including walnut, mahogany and teak" for different uses (Nicholas and Brown 2002). *A. melanoxylon* is a valued timber species known commercially by several common names: Australian blackwood, Tasmanian blackwood, *Acacia* blackwood, black sally, black wattle, lightwood, swarthout.

Blackwood is a highly decorative timber with a rich colour that is valued for high quality furniture, billiard tables, tool handles and boat building. The colour is medium golden, golden-brown, reddish brown and walnut-brown and contrasting bands of colour given by the growth rings may be identified

(Meier, 2015; Bradbury et al., 2010; Knapic, 2006; Igartúa, 2017). The wood colour may vary between and within trees (Zwaan, 1982; Nicholas and Brown, 2002). For instance, a study of heartwood colour in 196 trees from South Africa from 26 stands in four regions concluded that environment influenced darkness, uniformity and brown and grey pigmentation (Harrison, 1975).

The wood shows a uniform fine to medium texture given by a straight or slightly wavy grain and has a lustrous surface (Meier, 2015). It performs well regarding workability with hand and machine tools, can be screwed and nailed well, allows steam bending, and provides an excellent finish related to stains and polish although gluing properties are often variable (Meier, 2015). The sawing performance is comparable to that of other hardwoods (Haslett, 1986). The surface roughness is related to the cutting conditions and the rotation speed must be adjusted according to the wood density, especially in low density wood where low speed levels must be used (Aguilera and Zamora, 2009). Sapwood and heartwood present a distinct behaviour given by their different density. Some sawmill workers that manipulate blackwood may suffer of allergic contact dermatitis due to the presence of 2,6-dimethoxy-1,4-benzoquinone and acamelin (Hausen et al., 1990).

The timber is easy to dry, without major problems resulting from the low wood shrinkage and with no end splitting (Haslett, 1983). Blackwood can be kiln dried from green but this accentuates the variability of drying rates and therefore it is recommended that it should be air-dried to approximately 30% of moisture content and finished off by kiln drying (Haslett, 1983). The wood shows little internal collapse if air dried for several months before kiln drying (Darrow, 1995).

Concerning their anatomical and chemical properties, the *A. melanoxylon* wood has also good pulpwood potential giving acceptable pulp yields and paper properties (Anjos et al., 2011), as discussed in another chapter off this Book (Potential of *Acacia melanoxylon* for pulping).

The wood of *A. melanoxylon* is durable but it was reported to have low natural resistance to attack by fungi and other wood decay organisms e.g., untreated wood could decay in less than 5 years if in contact with the ground. The heartwood is reported to be resistant to preservative treatment while sapwood is moderately treatable.

The common uses of *A. melanoxylon* are for veneer, furniture, cabinetry, gunstocks, panelling, interior joinery, billiard tables, turned objects, cooperages, coachwork and boat building, sporting goods and other decorative wood objects (Meier, 2015; Searle, 2000). *A. melanoxylon* is often used in musical instruments like pianos and piano keys, violins and violin bows and xylophones as well as at the high end of the acoustic guitar market given its stability and acoustic qualities (Morrow, 2007; Bucur and Chivers, 1991, Meier, 2015). A recent study (Simanta et al., 2017) developed Glulam for exterior applications made with *Acacia* wood.

The potentials of using *Acacia* stands as fuel are not explored and as far as we know no studies about the use of Blackwood for this propose were made.

Biomass generated from controlling invasive *Acacias* species could be explored for bioenergy production but only in a perspective of control/eradication actions. In this case additional studies will be needed because the *Acacia* species are nitrogen fixing which bring the need to have equipment for energy recuperation with very specific in order not to produce NOx gases.

## ACKNOWLEDGMENTS

Centro de Estudos Florestais is a Research Unit funded by Fundacão para a Ciência e a Tecnologia within UID/AGR/UI00239/2013.

Institute for Sustainability and Innovation in Structural Engineering is financed by FEDER funds through the Competitiveness Factors Operational Program – COMPETE and by national funds through FCT - Fundação para a Ciência e a Tecnologia within the scope of the project POCI-01-0145-FEDER-007633.

## REFERENCES

Aguilera, A.; Zamora, R. *Eur J Wood Wood Prod.* 2009, 67(3), 297-301.

Anjos, O.; Santos, A.; Simões, R. *Appita J.* 2011, 64 (2), 185-191.

Arán, D.; García-Duro, J.; Reyes, O.; Casal, M. *Forest Ecol Manag,* 2013, 302, 7–13.

Bamber, R. K, Fukazawa, K. (1985). Sapwood and heartwood. A review. *Forest Abstracts*, 46:567-580.

Bamber, R. K. (1976). Heartwood, its function and formation. *Wood Science and Technology*, 10:1-8.

Barry, K. M.; Davies, N. W.; Mohammed, C. L.; Beadle, C. L. *Holzforschung.* 2003, 57, 230-236.

Bolza, E.; Kloot, N. H. The mechanical properties of 174 Australian timbers. *CSIRO Division of Forest Products Technical Paper* 1963. No. 25.

Bradbury, G. J.; Potts B. M.; Beadle C. L. *Forestry.* 2010, 83(2), 153-162.

Bradbury, G.; Potts, B.; Beadle, C. *Ann Forest Sci.* 2011, 68(8), 1363-1373.

Bucur, V.; Chivers, R. C. *Acta Acust United Ac.* 1991, 75(1), 69-74.

Carvalho A. *Madeiras portuguesas*; *Estrutura anatómica, propriedades, utilizações*; DGF: Lisboa, PT, 1977; Vol. II, 234-235. [*Portuguese woods*; *anatomical structure, properties, utilizations*, DGF: Lisboa, PT, 1977; Vol. II, 234-235].

Çetinkol, Ö. P.; Smith-Moritz, A. M.; Cheng, G.; Lao, J.; George, A; Hong, K.; Henry, R.; Simmons, B. A.; Heazlewood, J. L.; Holmes, B. M. *PLoS ONE.* 2012, 7(12), e52820.

Clark, N. B.; Balodis, V.; Fang, G.; Wang, J. In Australian Tree Species Research in China Pulpwood potential of *Acacias*; Brown, A. G. Ed., *Proceedings, Australian Centre for International Agricultural Research*, Canberra, Australian, 1992, pp 48.

Collins, D. J.; Pilotti, C. A.; Wallis, A. F. A. *Appita J.* 1990, 43,193–198.

Darrow, W. K. *Report on the Silviculture and Marketing of Blackwood (Acacia melanoxylon) and its Potential Role in South African Forestry.* Institute for Commercial Forestry Research Bulletin Series. Scottsville, South Africa. 1995, 3/95, pp 38.

Doran, J. C.; Gunn, B. V. In Australian *Acacias* in Developing Countries, Treatments to promote seed germination in Australian *Acacias*; Turnbull J. W. Ed. *Proceedings of an International Workshop*, Gympie, Qld., ACIAR Proceedings, Australia, 1987, 16, pp 57-63.

Duan, W.; Ohara, S.; Hashida, K.; Makino, R. *Holzforschung.* 2005, 59, 289–294.

Fernandes F. M.; Silva L.; Land E. O. *In Flora e fauna terrestre invasora na Macaronésia*; Silva L, Land E. O., Luengo J. L. R. Ed. Top 100 nos Açores, Madeira e Canárias. Arena, Ponta Delgada, 2008; pp. 342-345. [*Invasive terrestrial flora and fauna in Macaronesia*; Silva L, Land E. O., Luengo J. L. R. Ed. Top 100 nos Açores, Madeira e Canárias. Arena, Ponta Delgada, 2008; pp 342-345].

Freire, C. S. R.; Coelho, D. S. C.; Santos, N. M.; Silvestre, A. J. D.; Pascoal Neto, C. *Lipids.* 2005, 40(3), 317-322.

Freire, C. S. R.; Pinto, P. C. R.; Santiago A. S.; Silvestre A. J. D.; Evtuguin D. V.; Pascoal Neto C. *BioResources.* 2006, 1(1), 3-17.

Gourlay, I. D.; Barnes, R. B. *Common For Rev.* 1994, 73, 121–127.

Gourlay, I. D.; Kanowski, P. J. *IAWA J.* 1991, 12,187-194.

Harris, J. M.; Young, G. D. (1988). In *International Forestry Conference for the Australian Bicentennary, Wood properties of eucalypts and blackwood grown in New Zealand.* Albury-Wodonga, AFDI, New Zealand, Vol II, pp 8.

Harrison, C. M. *South For.* 1974, 15, 31 – 34.

Harrison, C. M. *South For.* 1975, 17, 49-56.

Haslett A. N. *N Z J For Sci.* 1983, 13(2), 130-8.

Haslett, A. N. Properties and utilisation of exotic specialty timbers grown in New Zealand. Part IV., Black walnut *Juglans nigra* L. Forest Research Institute, New Zealand Forest Service, Rotorua, N. Z 1986, pp 12.

Hausen, B. M.; Bruhn, G.; Tilsley, D. A. *Contact Dermititis,* 1990, 23, 33–39.

Hausen, B. M.; Schmalle, H. *Br J Ind Med,* 1981, 38, 105-109.

Hillis, H. E. *Heartwood and tree exudates*. Springer Verlag, Heidelberg, Berlin, London, Paris. 1987, pp 268.

Hoong, Y. B.; Paridah, M. T.; Luqman, C. A.; Koh, M. P.; Loh, Y. F. *Ind Crops Prod*. 2009, 30, 416–421.

Hoong, Y. B.; Pizzi, A.; Tahir, P. M. D.: Pasch, H. *Eur Polym J*. 2010, 46, 1268–1277.

Huang, Y., Wang, L., Chao, Y., Nawawi, D. S., Akiyama, T., Yokoyama, T., Matsumoto, Y. *J Wood Chem Technol*, 2016, 36(1), 9-15.

Igartúa, D. V.; Monteoliva, S. *Forest syst*, 2009, 18(1), 101-110.

Igartúa, D. V.; Moreno, K.; Monteoliva, S. E. *Forest syst,* 2017, 26(1), e007, 12 pages.

Ishida, Y.; Kitagawa, K.; Goto, K.; Ohtani, H. *J. Mass Spectrom*. 2005, 19, 706−710.

Jahan, M. S., Sabina, R., Rubaiyat, A. *Turk J Agric For.* 2008, 32, 339-347.

Jennings, S. M. *Aust Forestry*. 1998, 61, 141-146.

Jennings, S. M.: Hichey, J. E.; Candy, S. G. *Tasforests*. 2000, 12, 55–68.

Knapic, S.; Tavares, F.; Pereira, H. *Forestry*. 2006, 79, 371–380.

Le Floc'h, E. In *Biogeography of Mediterranean invasions, Invasive plants of the Mediterranean*; Groves R. H., Castri F. D. I. Ed; Cambridge University Press: Cambridge, UK, 1991; pp 67-80.

Lee, W. J.; Lan, W. C. *Bioresource technol*. 2006, 97(2), 257-264.

Leite, A.; Santos, C.; Saraiva, I.; Pinho, J. R. In *1° Encontro Invasoras Lenhosas*, Gerês, 16–18, novembro, Sociedade Portuguesa de Ciências Florestais, Lisbon, 1999; pp. 49–55. [*1st Meeting of lignocellulosic invasive species* Gerês, 16–18, novembro, Sociedade Portuguesa de Ciências Florestais, Lisbon, 1999; pp 49–55].

Lourenço, A.; Baptista, I; Gominho, J.; Pereira, H. *J Wood Sci*. 2008, 54(6), 464–469.

Lourenço, A.; Pereira, H. In *Lignin – Trends and Applications*. Chapter 3 Compositional variability of lignin in biomass; Matheus Poletto (Editor), InTech, Brasil, 2018; pp 65-98.

Luis, Â.; Gil, N.; Amaral, M. E.; Duarte, A. P. *Int J Pharmacol Pharm Sci*. 2012, 4(1), 225-231.

Machado, J.; Cruz, H. *Holz Roh- Werkst*. 2005, 63, 154-159.

Machado, J.; Louzada, J.; Santos, A.; Nunes, L.; Anjos, O.; Rodrigues, J.; Simões, R.; Pereira, H. *Mater Design*, 2014, 56, 975–980.

Makino, R.; Ohara, E.; Hashida, K. *J Trop For Sci*. 2009, 21(1), 45–49.

Mansilla, H.; García, R.; Tapia, J.; Durán, H.; Urzúa, S. *Wood Sci Technol*. 1991, 25, 145-149.

Maslin, B. R.; Pedley, L. *West Aust Herb res notes*. 1982a, 6, 1-128.

Maslin, B. R.; Pedley, L. *West Aust Herb res notes*. 1982b, 6,129-171.

Medhurst, J. L.; Pinkard, E. A.; Beadle, C. L.; Worledge, D. *Forest Ecol Manag*. 2003, 179,183-193.

Meier E. (2015). *Wood Identifying and Using Hundreds of Woods Worldwide*. www.wood-database.com.

Miller, J.; Murphy, D. In *XVII International Botanical Congress – Abstracts Many Acacia species*, Vienna, Austria, 2005, pp 10.

Miranda, I., Sousa, V., Pereira, H. *J Wood Sci,* 2011, 57(3), 171–8.

Monteoliva, S. E.; Iguartúa, D. V. *Revista Fac Agron Univ Nac La Plata*. [Agron Univ Nac La Plata magazine.] 2010, 109(1), 1-7.

Muhammad, A. J.; Ong, S. S.; Ratnam W. *J Forest Res*. 2018, 29(2), 549–555.

Nawawi, D. S.; Syafii, W.; Tomoda, I.; Uchida, Y.; Akiyama, T.; Yokoyama, T.; Matsumoto, Y. *J Wood Chem Technol*. 2017, 37(4), 273-282.

Neto, C. P.; Silvestre, A. J. D.; Evtuguin, D. V.; Freire, C. S. R.; Pinto, P. C. R.; Santiago, A. S.; Fardim, P.; Holmbom, B. *Nord Pulp Pap Res J*. 2004, 19, 513–520.

Nicholas, I.; Brown, I. *Forest Research Bulletin*. 2002, 225, pp 95.

Pietarinen, S.; Willför, S.; Holmbom, B. *Appita J*. 2004, 57(2), 146-150.

Pinto, I.; Pereira, H.; Usenius, A. *Trees*. 2004, 18, 284-294.

Pinto, P. C.; Evtuguin, D. V.; Neto, C. P. *Ind Eng Chem Res*. 2005a, 44, 9777-9784.

Pinto, P. C.; Evtuguin, D. V.; Neto, C. P. *J Agr Food Chem*. 2005b, 53, 7856-7862.

Quirk, J. T. *IAWA J*. 1983, 4 (2-3), 118-130.

Richardson, D. M.; Kluge, R. L. *Perspect Plant Ecol Syst*. 2008, 10, 161–177.

Rodrigues, C.; Santos, A.; Tavares, M.; Anjos, O. In *Proceedings of Wood Science and Engineering in the Third Millennium- ICWSE 2007*, Brasov, Romenia, 2007; pp 92-99.

Rucha, A.; Santos, A.; Campos, J.; Anjos, O.; Tavares, M. *Revista Floresta.* 2011, 41(1), 169-178.

Santos, A.; Anjos, O.; Amaral, M. E.; Gil, N.; Perreira, H.; Simões, R. *J Wood Sci.* 2012, 58, 479–486.

Santos, A.; Simões, A.; Tavares, M. *Forest Syst.* 2013, 22(3), 463-470.

Santos, A. J. A.; Anjos, O. M. S.; Simões, R. M. S. *Appita J.* 2006, 59(1), 58–64.

Santos, A. J. A.; Pereira, H.; Anjos, O. *Millenium.* 2018, 2(5), 13-19.

Schmalle, H. W.; Hausen, B. M. *Tetrahedron Lett.* 1980, 21 (2), 149-152.

Searle, S. D. *Aust For.* 2000, 63(2), 79-85.

Searle, S. D.; Owen, J. V. *Aust For.* 2005, 68(2),126-136.

Seigler, D. S. *Biochem Syst Ecol.* 2003, 31, 845-873.

Simanta, D.; Suryoatmono, B.; Tjondro, J. A. In *The 6th International Conference of Euro Asia Civil Engineering Forum (EACEF 2017)*, MATEC Web Conference, 2017, 138 -148.

Simes, J. J. H.; Tracey, J. G.; Webb, L. J.; Dunstan, W. J. (). An Australian phytochemical survey. III. Saponins in eastern Australian flowering plants. *CSIRO,* 1959, pp 31.

Soon, L. K.; Chiang, L. K. *Asian j appl sci.* 2012, 5(2), 107-116.

Tavares, F.; Louzada, J. L.; Pereira, H. *Eur J For Res.* 2014, 133, 31-39.

Tavares, F.; Quilhó, T.; Pereira, H. *Cerne.* 2011, 17(1), 61-68.

Tavares, M. *2º Relatório de execução material do Projecto POCTI/42594/AGR /2001 - Valorização do lenho de acácia produzido em Portugal. Potenciais utilizações.* FCT/INIAP Lisboa. 2004, pp 32 [*2nd progress Report of Project POCTI/42594/AGR/2001 - Valorisation of Acacia wood produced in Portugal. Potential uses.* FCT/INIAP Lisboa. 2004, pp 32].

Taylor, A. M.; Gartner, B. L.; Morrell, J. J. *Wood Fiber Sci.* 2002, 34, 587-611.

Tindale, M. D.; Roux, D. G. *Phytochemistry.* 1974, 13(5), 829-839.

Tomé, M.; Ribeiro, F.; Soares, P. (2001). *O modelo GLOBULUS 2.1. RTC-GIMREF nº1/2001*. Instituto Superior de Agronomia - Departamento de Engenharia Florestal. Lisboa. [*The GLOBULUS 2.1 model. RTC-GIMREF nº1/2001*.].

Wei, S. D.; Zhou, H. C.; Lin, Y. M.; Liao, M. M.; Chai, W. M. *Molecules*. 2010, 15, 4369−4381.

Whibley, D. J. E. and Symon, D. E. *Acacias of South Australia*, Ed.2. Government Printer: Adelaide, South Australia, 1992, pp 328.

Wilkins, A. P.; Papassotiriou, S. *IAWA J*. 1989, 10(2), 201-207.

Willför, S.; Sundberg, A.; Pranovich, A.; Holmbom, B. *Wood Sci Technol*. 2005, 39, 601–617.

Zhang, S. Y. In *Proc Timber Management Toward Wood Quality and End-Product Value*. CTIA/IUFRO International Wood Quality Workshop. Part I. 1997; pp. 22-28.

Zwaan, J. G. *South Afr For J*. 1982, 121, 38-43.


*Reviewed by*: José Luis Lousada, CITAB – Centro de Investigação e Tecnologias Agroambientais e Biológicas, Vila Real. UTAD – Universidade de Trás-os-Montes e Alto Douro, Departamento de Ciências Florestais e Arquitetura Paisagista, Vila Real.

In: *Acacia*
Editor: Aide Matheson

ISBN: 978-1-53614-237-2
© 2018 Nova Science Publishers, Inc.

*Chapter 2*

# ECOLOGY OF ACACIEAE

## *Marcos S. Karlin[1],\* and Ulf O. Karlin[2]*

[1]Universidad Nacional de Córdoba, Facultad de Ciencias
Agropecuarias, Department of Natural Resources, Córdoba, Argentina
[2]Universidad Nacional de Chilecito, Chilecito, La Rioja, Argentina

### ABSTRACT

In 2005, the *Acacia* genus was proposed to be split into several genera, triggering a heated discussion between the botanists of Australia, Africa, and America. This re-typification was ratified in 2006 at the Vienna's Nomenclature Session of the 17th International Botanical Congress (IBC), and later in 2011 in the Melbourne's Nomenclature Section of the 18th IBC. The decisions taken in Melbourne confirm the Australian *Acacias* retain the name, while new names are needed for the African and American species of the Acacieae tribe. Currently, the tribe is formed by the genera *Acacia* (for Australian species), *Vachellia*, *Senegalia*, *Parasenegalia*, *Acaciella*, *Mariosousa*, and *Racosperma*. Despite this division, the species of the tribe along several ecosystems around the globe perform similar characteristics and behaviors. Most species seem to have co-evolved with frequent forest fire events and developed the ability to resprout or activate germination after a fire event. Some *Acacia sensu lato* (*s.l.*) species have

\* Corresponding Author Email: mkarlin@agro.unc.edu.ar.

also co-evolved with wildlife and seem to adopt resistance to defoliation of seedling browsing. It is well-known that *Acacia s.l.* species have the ability to fix atmospheric nitrogen and, coupled with important leaf litter inputs to the soil, they may modify soil chemistry and physics by enabling microorganisms and soil fauna to alter the microhabitat beneath the tree. Such characteristics are significant in the recovery of ecosystems within ecological successions after disturbance, but may also alter original ecosystems when *Acacias* invade habitats of other native species. These characteristics and others have been found in several species of the tribe Acacieae worldwide. The objective of the chapter is to describe and analyze common characteristics among *Acacia s.l.* species and to trace some parallelisms of their performance throughout several ecosystems that hold such species.

**Keywords**: *Acacia*, *Vachellia*, *Senegalia*, evolution, relationships, fire, invasiveness

# INTRODUCTION[1]

In 2005, the *Acacia* genus was proposed to be split into several genera, triggering a heated discussion between the botanists of Australia, Africa, and America. This re-typification was ratified in 2006 at the Vienna's Nomenclature Session of the 17th International Botanical Congress (IBC) (Maslin, 2008), and later in 2011 in the Melbourne's Nomenclature Section of the 18th IBC (Thiele et al., 2011; Moore et al., 2011). The decisions taken in Melbourne confirm the Australian *Acacias* retain the name, while new names are needed for the African and American species of the Acacieae tribe. Currently, the tribe is formed by the genera *Acacia* (*sensu stricto –s.s.*, for the majority of the Australian species), *Vachellia*, *Senegalia*, *Parasenegalia*, *Pseudosenegalia*, *Acaciella*, and *Mariosousa*.

Despite this division, the species of the tribe along several ecosystems around the globe perform similar characteristics and behaviors. Most

---

[1] Note for the reader: along the chapter, the authors will employ the terms "Acacieae," "*Acacia s.l.*" or simply "Acacias" as generalizations of the analyzed genera of *Vachellia*, *Senegalia*, *Parasenegalia*, *Pseudosenegalia*, *Acaciella*, and *Mariosousa*. When the term "*Acacia*" is cited, it will refer specifically to the genus sensu stricto. Scientific names were updated at May 2018, according to the Catalogue of Life and Tropicos.org databases.

species, especially those from Africa and America, have evolved under the influence of herbivores by producing spines or secondary metabolites, or even interacting with other organisms like ants. Most species seem to have adapted to frequent forest fire events, by developing the ability to resprout or by eliminating seed dormancy mechanisms.

It is well-known that *Acacia s.l.* species have the ability to fix atmospheric nitrogen and, coupled with important leaf litter inputs to the soil, they may modify soil chemistry and physics by enabling microorganisms and soil fauna to alter the microhabitat beneath the tree. Such characteristics are significant in the recovery of ecosystems within ecological successions after disturbance, but may also alter original ecosystems when *Acacias* invade habitats of other native species.

The objective of the chapter is to describe and analyze common and differential characteristics among *Acacia s.l.* species and to trace some similarities of their performance throughout several ecosystems that hold such species.

Loads of information have been published worldwide about Acacieae, but there are no publications that have synthesized and systematized in one place the main facts of the ecological behavior of the species of this tribe.

The information incorporated in this chapter may be useful for the adoption of management or investigation policies not only for Acacieae but also for ecologically similar species like *Prosopis* or *Mimosa*, among other.

In this chapter there will be described ecological aspects such as climatic and edaphic factors conditioning the performance and distribution of *Acacias* worldwide, from species adapted to hyper-arid to tropical ecosystems, inquiring about their origins and their evolution along the continents that hold them. In addition, ecological relations with other organisms (plants, animals, bacteria) will be analyzed, trying to relate their natural behavior to possible lines of management for economic purposes or conservation. In relation to fire events, ecological adaptations and responses will also be analyzed, integrating other ecological factors that might alter *Acacias* population dynamics. Finally, *Acacias* as alien invasive species can alter ecological networks; a focus on the modification of the structure and dynamics of affected natural ecosystems will be made.

## CLIMATE AND SOIL

**Paleoclimate and Expansion**

There has been a convergent evolution within the paraphyletic genus as species invaded and dominated new arid landscapes that expanded at that time (Miller & Burd, 2014). They have evolved into three main genera (*Senegalia*, *Vachellia*, and *Acacia*) and four minor genera (*Acaciella*, *Mariosousa*, *Parasenegalia*, and *Pseudosenegalia*).

*Acacia s.l.* species, generally unarmed, are distributed throughout tropical and subtropical regions in the world (Figure 1), occupying arid and semiarid areas and a diverse spectrum of soils. The latitudinal limits are the 37°N and 42°S parallels.

*Acacia s.s.* genus is currently distributed over Oceania, being Australia the country with the largest richness of species. *Vachellia*, with spinescent stipules, shows the largest number of species in Mexico and eastern Africa. *Senegalia*, with recurved prickles on the stems, is more widely distributed around the world, coring large amount of species in Brazil, Bolivia or Mexico, and secondarily in Peru and eastern Africa (World Wide Wattle, updated March 2018). Minor genera like *Acaciella* and *Mariosousa* have the main distribution core in Mexico, while *Parasenegalia* distributes in Central and Southern America, and *Pseudosenegalia* can only be found in Bolivia (Seigler et al., 2017).

Most *Acacia s.l.* species are phenotypically plastic and may develop in a broad range of environments, in habitats with variable rainfall and altitude ranges; they are drought tolerant and can occur even in extremely arid conditions. Despite that *Acacia s.l.* occur most usually in dry regions with scarce and markedly seasonal rainfall, it can also be found in rainy tropical regions.

Their xerophytic morphological characteristics suggest they might have evolved in arid and semi-arid environments; however, Ross (1981) considers probable that *Acacia s.l.* species might have emerged in tropical lowlands and that most of the xerotypic features are secondary.

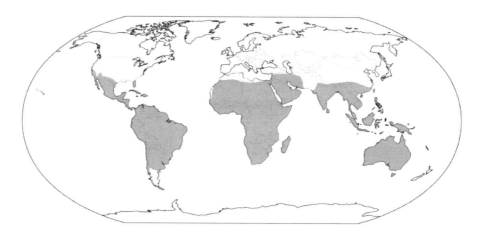

Figure 1. *Acacia s.l.* natural distribution map. Based on World Wide Wattle distribution maps (http://worldwidewattle.com), GBIF (http://gbif.org) and Ross (1981).

It is suggested that Acacieae might have evolved from their ancestors with morphological characters similar to the current *Acaciella* genus (Guinet & Vassal, 1978). Then, these ancestors might have spread throughout Gondwanaland sometime between the early Cretaceous and the Paleocene, during a warmer period, between 140 and 60 Mya.

During the Paleogene, Acacieae species might have spread over rainforests stretched through current desertic regions like southern South America, northern Africa, and Australia. Since Acacieae species are low-light intolerant, their ancestors' abundance might not have been important within rainforests, and this hypothesis is supported by the lack of fossil pollen records. It is also hypothesized that these species might have been climbers and lianes in order to seek for light, with similar characters as some species of the genus *Senegalia* (Ross, 1981).

Pollen fossils suggest the occurrence of Acacieae phylogenetic differentiation around late Oligocene-early Miocene (~23 million years ago) (Mildenhall, 1972; Bouchenak-Khelladi et al., 2010), during a period at least 3°C warmer than current air mid temperatures, and through a period of [$CO_2$] reduction from 600 to 300 ppm. Throughout this climatic transition, seemingly C3 grasses gradually began to expand over open woodlands and savannas (Kürschner et al., 2008). Such changing –fire-prone– climatic conditions might have altered tree saplings growth and promoted the expansion of grasses (Bond et al., 2003). Grasslands might have expanded, and perhaps among them, Acacieae species did too.

After the appearance of Acacieae, diversification, hybridization and introgression allowed species to adapt to changing conditions (in general terms, colder and more arid conditions) and to expand to different habitats throughout the Plio-Pleistocene (~5 Mya - ~10 Kya) (Odee et al., 2012), maybe accompanied by the appearance and expansion of C4 fire-prone grasses.

Acacieae may have expanded and colonized several regions in the world, depending on climate shifts along recent geological history. Their ability to disseminate pollen and seeds over wide distances by biotic and abiotic vehicles enables Acacieae species to expand throughout large areas.

*Acacias* tend to be dominant in low forests, woodlands or shrublands, but they can be co-dominant in grasslands across the world. It is an iconical feature in African savannas, in central Argentina highlands, or in Australian shrublands, known as mulgas (*A. aneura*, "mulga" gives its name to this semi-desertic environment).

## Climatic Adaptation

*Acacia s.l.* distribution is thought to be driven by climate, especially by rainfall, more than temperature. *Acacias* cover a wide range of isohyets; the lower limit seemingly depends on the type of soil, being of 200-250 mm for sandy soils and 400 mm for clayey soils (Obeid & El Din, 1971).

However, *Vachellia tortilis* and *Acacia ehrenbergiana* can grow isolatedly in the wadis of the hyperarid deserts of Saudi Arabia and Sahara with less than 50 mm year$^{-1}$ of rainfall, as recorded for the surroundings of Al-Madinah, Saudi Arabia (Al-Mefarrej, 2012) or for Fezzan region, Libya (El-Bana & Al-Mathnani, 2009). *V. tortillis* may adopt unusual architecture in the southeastern Arabian peninsula (~25ºN), under hyperarid conditions, showing a southerly tilted canopy (Figure 2) as a possible adaptation for sun orientation, maximizing sunlight interception during the hottest hours of mid-day and reducing soil evaporation by reducing in 12°C the soil temperature (Ross & Burt, 2015).

Figure 2. *Vachellia tortilis* tilted canopy, UAE. Creative Commons.

Apparently, Acacieae have no upper limits regarding the amount of rainfall they can receive. Some genera like *Senegalia* in Brazil grow naturally in rainforests with up to 1700 mm year$^{-1}$, e.g., *Senegalia tenuifolia* in the Atlantic rainforest, southeastern of Brazil (Brandes & Barros, 2007; Ferreira Barros & Pires Morim, 2014). Some alien Acacieae species can also spread through tropical areas, e.g., *Vachellia nilotica* in tropical areas of Australia, Indonesia, Vietnam, with rainfalls up to 1500 mm year$^{-1}$ (Kriticos et al., 2003; Rivers, 2017). Some *Acacia* species have been cultivated in tropical areas, as happens with *A. mangium* in logged tropical forests of Borneo (up to 3300 mm year$^{-1}$) as a fast-growing introduced species. Nevertheless, in tropical areas, *Acacias* have limited growth depending on the amount of light they receive, and therefore, they need to grow in cleared highlighted areas.

Regarding the thermal limits of the Acacieae, since these have a pantropical distribution, they mostly adapt to elevated temperatures. However, some species can develop in temperate or even cold conditions.

In the Florentine Valley, Tasmania (42°30'S), *Acacias* (*A. dealbata* and *A. melanoxylon*) can grow under the canopy of temperate mixed rainforests, and the region can reach minimum mid temperatures in the coldest month of about -0.5°C (Gilbert, 1959; Dean et al., 2011).

*Vachellia caven* is known to be widespread over the Chaco Árido, the Chaco Serrano and the Espinal ecoregion in Argentina, and between the III and the VIII Region in Chile (Aronson, 1992). Latitudinal limits reach the 33°S within the territory of Argentina, however, *V. caven* can be seen isolatedly in the Monte region reaching Northeastern Chubut (Argentina) at 43°S (Encyclopedia of Life, accessed April 2018), with medium annual temperatures of 13.6°C and minimum mid temperatures in the coldest month of about 2.1°C. In Chile, it reaches the 37°S, with medium annual temperatures of 13°C and minimum mid temperatures of the coldest month of about 4.9°C.

*Senegalia greggii* occurs along the border limits of Mexico and US, over the dry-washes of local deserts. It is cited as the most meridional widespread *Acacia s.l.* species, reaching the southwestern extreme of the Utah State (37°10' N), in the Mojave Desert (Gaddis, 2014), with medium annual temperatures of 16.1°C and minimum mid temperatures in the coldest month of about -3.5°C.

## Soil Relations

The distribution of *Acacias* throughout the world suggests a wide adaptation of these species for several soil features. Climate, as one of the factors of soil formation, contributes modifying soil characters. As *Acacias* distribute all over the world throughout climatic extremes, it suggests also that *Acacias* should develop over most of the soil orders.

According to what has been described in the subsection *Climatic Adaptation*, *Acacias* can grow over sandy dunes (e.g., Entisols) in desert regions, over clayey soils in tropical rainforests (e.g., Oxisols), over alkaline soils (e.g., Aridisols) or over acidic soils (e.g., Spodosols).

During the phylogenetic differentiation, *Acacia s.l.* populations have spread over specific niches according to soil characteristics. It is suggested that geochemical gradients may play a driving role in the diversification of Acacieae (Bui et al., 2014).

For instance, in Australia, subsoil pH, $Na^+$, and EC are all-important predictors of the distribution of *Acacias* in the SW bioregion. Apparently, *Acacias* have co-evolved with landscape changes in Australia; changes in pedogenesis since the Miocene, such as salinization or calcification in arid zones, matches the period when radiation and differentiation of *Acacias* occurred (Bui et al., 2014) (See subsection *Paleoclimate and Expansion*).

Some Acacieae species induce simultaneous changes in the above and below ground communities, microclimates, soil moisture regimes, and soil nutrient levels. *Acacias* may perform as important dune stabilizers in arid lands and coastal ecosystems and may enable important changes as soil transformer. However, after the introduction of Australian *Acacias* into foreign countries like South Africa or Portugal, they have become a threat to local biodiversity (Avis, 1989).

Most tropical or subtropical soils have low N contents, but they can be adequate niches for *Acacia s.l.* species since they have the ability to nodulate and fix atmospheric N beyond their physiological needs. The excess of nitrogen gained is finally converted into nitrates, which can be used by other species, modifying occupied niches, soil-plant relations and successional dynamics (Kuyper & de Goede, 2013).

*Acacia* trees may contribute to the development of fertile islands in arid lands. Scattered trees in arid lands can offer food, nest sites or shade to wildlife and livestock, increasing the deposition of feces or material from fallen nests. Dean et al. (1999) found soil concentrations of N, K, and P two to three times greater under the canopies of *Vachellia erioloba* than in the interpatches, in the Kalahari Desert.

As cultivated in agroforestry systems, some *Acacias* can contribute with litter into the soil. Under tropical climates, *Acacia mangium* has fast-growing rates and liberates large amounts of litter with C/N relations close to 35:1, with the consequently increasing amounts of organic matter. Nitrogen has variable rates of liberation according to specific site conditions, depending on moisture, temperature, microbial activity, immobilization processes and understory vegetation; nevertheless, litter contribution increases soil N compared with control areas without litter contribution (Xiong et al., 2008; Castellanos-Barliza & León Peláez, 2011).

Phosphorous has different dynamics according to the characteristics of soils; available organic phosphorous is usually increased with litter contribution (Xiong et al., 2008), however, it usually becomes immobilized in naturally acidic ferralitic soils or anthropically acidified soils (Castellanos-Barliza & León Peláez, 2011). If litter accumulation becomes important in *Acacia* plantations, pH decreases even more respect soil control (Xiong et al., 2008), and nutrients may become unavailable.

In alkaline soils, *Acacias* can also modify soil quality in agroforestry systems. Soil organic carbon and nitrogen improved in the alkaline soils of the northern semi-arid India that were cultivated with *Vachellia nilotica* alone or in arrangements with crop rotations of *Oryza sativa* and *Trifolium alexandrium*. The last increased 40 to 70% the amounts of total nitrogen in soil than for cropping alone. Regarding N mineralization rates, total N release was the highest in the *Vachellia* + *Oryza-Trifolium* system (312.90 kg N ha$^{-1}$) followed by *Vachellia* alone (297.14 kg N ha$^{-1}$) (Kaur et al., 2000).

## PLANT-PLANT RELATIONSHIPS

*Acacias* may perform as competitors or facilitators depending on the environmental characteristics. For arid and semi-arid conditions, water is usually the limiting resource, and canopy usually facilitates understory growth under drier conditions (Dohn et al., 2013). When rainfall becomes less restrictive, other resources rather than water become restrictive. In tropical forests, the limiting conditions are usually related to light or nutrient availability (Chou et al., 2017).

Under semi-arid conditions, *Acacias* tend to encroach easily due to its ability to compete with other species. While water becomes less restrictive within the ecosystem, *Acacias* switch from facilitators to competitors, especially with grasses, restricting the forage supply of livestock. It also reduces cattle circulation, reduces air circulation and increases ectoparasite proliferation.

Dohn et al. (2013) suggests that a shift occurs from net competitive to net facilitative effects of trees on understory grass production with decreasing annual precipitation in savanna ecosystems. Trees may facilitate grass growth, improving soil water availability related to hydraulic lift from deep-rooted trees or shrubs, or through a reduction of evapotranspiration in the understory.

Under the canopy of *Acacias* total biomass, biodiversity, grass and forbs biomass might differ with the amount of rainfall. With 500 mm yr$^{-1}$ in Ethiopian savannas, biomass, biodiversity and forbs biomass (with a higher shading tolerance than grasses) tend to be higher than in the intercanopy. Grasses differ in its behavior depending on the density of the canopy; for instance, under *Vachellia tortilis* canopy grasses reduce their biomass production and increase it in the intercanopy (Linstädter et al., 2016) possibly due to shading effects but also due to water competition with trees. Under other *Vachellia* species, grasses tend to be insensitive to canopy effect. In central Argentina, with 800 mm yr$^{-1}$ and under secondary *Acacia* forests, grasses reduce their biomass production with *Acacias* shading while forbs biomass and biodiversity has no effect with changing tree canopies (Karlin et al., unpublished data).

In studies carried out in Tanzania, Ludwig et al. (2003 and 2004) discovered that *Vachellia tortilis* can lift-up and exude around 75 to 225 liters of water each night to an area of 300 m$^2$. However, these authors suggest that when the density of *Vachellia* roots is high in the topsoil competition with understory vegetation probably prevails, despite the fact that there might be an effect of hydraulic lift and understory plants may receive some benefits from it.

In semi-arid savannas of southern Ethiopia, Linstädter et al. (2016) found that canopy density and encroachment are inversely related, likely due to an effect of intraspecific competition for soil water. Perhaps denser canopies like in *V. tortilis* have higher hydric demand and each individual tree will therefore have a higher threshold for its nearer neighbor. The opposite was found for *V. drepanolobium*; with half the canopy density than *V. tortilis*, it shows more tendency to encroach.

The canopy can equally affect the development of saplings. Since Acacieae are usually light-demanding there can be expected that shading may affect sapling growth. Venier et al. (2012) found that the species of the genus *Senegalia* (*S. gilliesii* and *S. praecox*) are, on average, more insensitive to light availability than *Vachellia* (*V. aroma*, *V. caven*, *V. astringens*). Since the last three usually form secondary forests with open canopies, saplings usually do not receive shading effects, while *Senegalia* species are usually accompanying species of climax forests of *Aspidosperma quebracho-blanco* (Karlin et al., 2013) or *Schinopsis marginata* (Karlin et al., 2015), and therefore might be adapted to shade.

Seemingly, some plant-plant interactions may indirectly favor ant-*Acacia* mutualism. Palmer et al. (2017) inform that the ant-host *Vachellia drepanolobium* reduces soil moisture losses under the influence of neighbor grasses and indirectly facilitates the survival of *Crematogaster mimosae* colonies by inducing higher nectaries activity.

Under arid conditions, nurse plants usually increase soil water content and facilitate the growth of understory plants. In the Kalahari Desert, with about 200 mm year$^{-1}$ of rainfall, *Vachellia erioloba* performs as an important nurse plant. Many species can only grow under *Vachellia* or other trees due to the lower diurnal temperature fluctuations respect open spaces. Many species occurring under *Vachellia* canopy do not germinate outside the shade influence (Kos & Poschlod, 2007). Nevertheless, if nurse *Acacias* produce large amounts of litter understory, it may contribute to a drastic reduction of light and might be deleterious for understory species, despite the higher amounts of soil water under the nurse plants influence (Helman et al., 2017).

In tropical forests, some *Acacias* can grow as lianas and may compete strongly with understory plants. Lianas usually have the function of cover rapidly forest gaps and this way regulating the development of ecological succession, competing with light, water, and nutrients. There are reports that *Senegalia kamerunensis* competes for above and below-ground resources with understory plants in tropical forests, especially with pioneer species such as *Nauclea diderrichii*, affecting relative growth rates (Toledo-Aceves & Swaine, 2008).

## PLANT-ANIMAL RELATIONSHIPS

### Livestock and Wildlife Ecology

In several regions of the world, *Acacia s.l.* species constitute a key source of fodder for livestock (cows, goats, camels, etc.).

In adult *Acacia* plants, branches can be browsed by herbivores, who ingest protein-rich leaves. When branches are browsed from African species like *Vachellia drepanolobium*, *V. nilotica* or *V. tortilis*, plants may respond by forming denser and longer spines (Young & Okello, 1998; Brooks & Owen-Smith, 1994; Rohner & Ward, 1997). Branches above the reach of herbivores are not browsed and these produce much shorter spines (Young, 1987; Young & Okello, 1998). It seems to represent a physiological mechanism for defense, rather than genetic, in which early herbivorism induce a higher rate of spinescence and consequently increases plant fitness. Apparently, this energetically wasteful mechanism may reduce the ability to induce reproductive mechanisms (Karban et al., 1999).

Spine production is not only energetically wasteful, but it might also be nutrient limited. It is suggested that *Vachellia tortilis* produce longer spines under nutrient-rich soil conditions. High soil fertility induces higher plant growth and forage availability for herbivores; nevertheless, a larger spine production may be also induced as a defense mechanism (Gowda et al., 2003).

South American Acacieae have also developed spinescence, and it tends to increase under strong grazing pressure, despite that the density and size of the wild herbivores are not as large as the Africans are. In the Chaco region, Acacieae seems to have co-evolved during the Pleistocene with larger herbivores like Glyptodontidae, Megatheriidae, Mylodontidae, Machrauchenidae, Toxodontidae, Gomphoteridae, and Equiidae. After Pleistocene, the megafauna has shifted to minor herbivores like camelids, small deers or tapirs, spread along large grazing or browsing areas. Seemingly, native ants have occupied the place of megafauna as the main herbivores in this subtropical environment (Bucher, 1987). After the arrival of European colonizers, livestock became one of the main factors affecting vegetation physiognomy and altering natural fire cycles. Under this new context, plants apparently have been negatively selected due to antiherbivore traits, by browsing rejection due to spinescence or the production of secondary compounds. Acacieae species might have been some of those that have adapted to this new scenario.

On the contrary, the Australian *Acacias* (except for small shrubs in open arid areas such as *A. spinescens*, *A. colletioides*, *A. acanthoclada*, and *A. ferocior*) have lost the stipular spines, perhaps because of the lack of large natural browsers in this continent (Brown, 1960; Bucher, 1987). Brown (1960) suggests that spinescence in the armed Australian members of the genus has arisen secondarily through the modification of phyllodes, branchlets, or even peduncles, passing through an earlier period in which spininess had little adaptive value, followed by a time in which selective pressure again arose favoring the development of spines *de novo*. During the Plio-Pleistocene the bigger known mammals were the giant marsupials *Macropus*, *Diprotodon*, and *Nototherium*, who seem to have been better grazers than browsers.

As spinescence is clearly increased proportionally by herbivory, the induction of secondary compounds is seemingly activated only under specific conditions. Secondary compounds like tannins, saponins, oxalates, hydrogen cyanide, mimosine, among others, seem to be produced in *Acacias* when browsing is relatively intense or when environmental conditions are harsh (Rohner & Ward, 1997; Dynes & Schlink, 2002; Rossi et al., 2007; Pedraza Olivera, 2008). Nevertheless, it is not possible to assert direct relations between herbivory intensity and the concentration of secondary compounds as a defense response in Acacieae, in different ecological or phenological conditions, due to contradictory or inconclusive results among several studies (Brooks & Owen-Smith, 1994; Gowda, 1997; Scogings et al., 2015; Wigley et al., 2015).

It is quite ordinary for saplings to be consumed by livestock when loads are high or when the grasses have low palatability. Wildlife in African savannas can also browse saplings, especially in areas with low grass cover, where predators cannot lurk and herbivores can escape (Barnes, 2001). This consumption can alter natural regeneration or retard the growth of saplings in *Acacias*, subtracting possibilities of survival for drought or fire. Sapling browsing is usually intensified when these are associated with water sources in arid and semiarid regions.

Saplings are most commonly consumed before they become lignified or gain a determined height. In Botswana, elephants browse saplings below 1 m high (Barnes, 2001).

The physiognomy of *Acacia* populations is often related to saplings development. When browsing eliminates apical buds, a coppice growth is typically produced and plants grow multistemmed.

## Ant-*Acacia* Mutualism

The plants performing as a natural habitat for ants are called myrmecophytes. Ants may use hollow stems, twigs, thorns or any other cavity of the plant as a shelter for nests. Plants also usually carry extrafloral nectaries, leaf surface pearl glands, Beltian bodies, trichilia or other organs furnishing concentrated sugary, fatty or proteinaceous matter utilized as food by the guest ant (Brown, 1960).

True myrmecophyte *Acacias* can be seen in The Americas and Africa, but not in Australia due to the lack of spines (Brown, 1960). Ants host

*Acacia* plants with swollen spines. Ant hosting, as happens with spininess, seems to represent an ecological adaptation for herbivores attack, that is evident elsewhere but in Australia. Ants can also protect their host from phytophage insects or competing plants. In Mesoamerica, *Pseudomyrmex/ Vachellia* mutualism may have strengthened in the late Miocene (~10-12 Mya), during a shift from closed to open and dry habitats, where browsers became a significant selective force on local plants (Ward & Branstetter, 2017).

In Kenya, *Crematogaster* spp. ants colonize *Vachellia drepanolobium* plants. Ants depend on productivity gradients spatially correlated with termite mounds, where the density of litter-dwelling invertebrates is increased. These invertebrates represent a food source for *Crematogaster* ants. Related to productivity gradients, *Vachellia* plants grow faster in areas with higher density of mounds, where fertility and water content are increased. Subordinate ant species are usually replaced by dominant ant species over the hosts with higher growth rates, while the opposite occurs when plants have slower growth rates. It is suggested that dominant ants are nitrogen-dependent because of their protein-rich exoskeleton, and therefore they have success in high mound density environments. In contrast, subordinate ants seem to be more stress-resistant in poorer environments (Palmer, 2003) and can apply survival strategies in disturbed areas like fire-affected stands, presumably evacuating sooner and rapidly than dominant ants (Sensenig et al., 2017).

*V. drepanolobium* has shorter spines than other Acacieae species without symbiotic ants (e.g., *V. seyal*). This pattern may indicate a trade-off between the production of ant attractors and spines as defense strategies against browsing (Madden & Young, 1992).

Ants protect young flower buds from phytophages by increasing the number of visits, but by its absence, during peak flower fertility, allow access to pollinators. After pollination, visits increase again to deter beetles and seed predators. This way, it is possible to correlate the number of visits with the number of fruit-set per inflorescence. Willmer & Stone (1997) demonstrate that for the mutualism between *Crematogaster* and *V. drepanolobium* or *V. zanzibarica* in Tanzania, the amount of fruit-set obtained per inflorescence increases by the unit per every 30 ants found in a 50-cm branch section.

A similar behavior is reported in the Mesoamerican ant-*Acacia* mutualism between *Pseudomyrmex veneficus* and *Vachellia hindsii* (Raine et al., 2002). Plants synchronize pollen and nectar production respectively, with pollinators and ants; pollen is liberated during the first hours of the day (6:00 to 10:00); next, pollinators visit the flowers (8:00 to 13:00); subsequently nectar is released (10:00 to 14:00) and ants are held busy collecting it while pollinators cease to collect pollen. From 10:00 to 20:00 (with peaks between 13:00 and 17:00) ants visit the Beltian bodies where they collect their protein source. This mechanism shows how pollinators have a window for pollination, and afterward, flowers receive protection from aggressive mutualistic ants.

Both examples reveal incredible similarities in convergent evolutionary processes occurring in different geographic locations, meaning that Acacieae species (*Vachellia* in both cases) reflect similar functional characters.

## Seed Dispersal by Animals

Livestock and wildlife are important seed dispersers of Acacieae species. In Acacieae seeds, the epidermis are thickened forming an impermeable hard coat. A seed hard coat represent a physiological adaptation that performs as a barrier to water, gases or microorganisms and protects the embryo for very long periods after seed production by inducing physical dormancy.

When Acacieae fruits are consumed by ungulates, they may soften seed hard coats and may eliminate dormancy, enabling germination after seed dissemination by feces deposition. However, not every Acacieae species respond positively to this treatment; Venier et al. (2012) inform that *Acacia s.l.* species in Central Argentina have differential responses by the passing of seeds through the digestive tract of ruminants, depending on the thickness of the seed hard coat. *Vachellia astringens*, *V. caven* and *V. aroma* differ in the epidermis and sclerified parenchyma thickness; *V. aroma* has the thinnest coat and represent the unique species of the three that show a positive effect to a simulated ruminal + acid scarification treatment by activating germination (up to 70%). *V. astringens* and *V. caven* showed no effect on germination when they were treated.

Endozoochory effect has also been registered in Asia and Africa. Germination rates increase in Acacieae species when these pass through ungulates' guts. Results from several authors have been compiled by Or & Ward (2003) regarding the effect of seed consumption of African and Middle East *Acacias* (several subspecies of *Vachellia tortilis* and *V. nilotica*) by gazelles, kudus, giraffes, elephants, antelopes or ostriches.

In the Middle East, endozoochory seems to represent a compensating mechanism in the regulation of *Vachellia tortilis* populations. In grazed areas, ungulates remove the pods that fall to the ground disseminating the seeds by feces and increasing sapling frequency. The distance of dispersion depends on the time the seeds stay in the digestive tract of the animal; usually, the dispersal distance is directly related to the body size of the disperser. Seeds are usually deposited in open areas rather than in understory; therefore, new saplings prevent interspecific competition (Or & Ward, 2003) but they may be exposed to unfavorable ecological conditions.

Nevertheless, when browsing pressure is high, tree height distribution is altered by reducing the survival and frequency of taller saplings, in contrast to unbrowsed areas where height frequencies are better balanced. If grazing intensity is low, pods accumulate under the trees and these are then affected by bruchids, reducing the seed viability (Rohner & Ward, 1999). Ungulates, therefore, perform as important system regulators by disseminating seeds, by activating germination and by indirectly reducing bruchids attack. If later browsing is managed and *Vachellia* populations are controlled, stands may increase in size and density (Figure 3).

*Acacia* seeds can exhibit morphological differences among species depending on the main vector of dissemination. Australian *Acacias* show adaptations for the ant or bird dispersal (Davidson & Morton, 1984). Probable myrmecochores are characterized by seeds with small and white arils, with relatively low energy reward. Apparently, ants are attracted to white colorations instead of a black or dark background (Whitney & Stanton, 2004). On the contrary, seeds containing larger and colored arils are eaten by birds. Birds are mainly attracted by reds and yellows. An example of morphological differentiation can be seen in *Acacia ligulata*, in which populations differ according to geographical location, and therefore to ecological conditions and possibly to vector availability.

Endozoochory has also been registered as a diffusion strategy for alien *Acacias*. Such is the case of the American *Vachellia farnesiana* over the Gran Canaria Island when seeds are consumed by rabbits. Despite that germination rates do not improve by passing through rabbits' digestive tract, it seems germination speeds up accelerating it. Seeds passing through the rabbit gut, possibly need a lesser investment of water to germinate than the non-ingested seeds. This situation may enable *V. farnesiana* to colonize dryer territories and therefore extend the invaded area (Pascual et al., 2009).

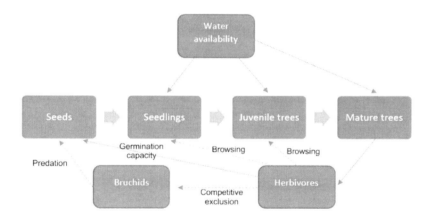

Figure 3. *Vachellia*-bruchids-ungulates-water interactions in Middle East ecosystems. (Adapted from Rohner & Ward, 1999). Full arrows indicate positive interactions. Broken arrows indicate negative interactions. Reproduced with the author's permission.

## FIRE ECOLOGY

*Acacia s.l.* communities usually comprise fire-prone ecological systems due to the tree architecture, canopy density, and connectivity. When dominant, *Acacias* form open shrublands that enable light to reach underbrush, favoring the development of grasses. Depending on soil conditions (especially nitrogen content), rainfall and grazing intensity, grasses may produce critical loads of combustible biomass (Figure 4). Understory shrubs (e.g., *Aloysia gratissima* in central Argentina) may also promote forest fires and may even increase fire intensity compared with grasses.

Nevertheless, there are *Acacia* habitats (e.g., mulgas in Australia) where the vegetation is distributed in patches, according to the "source-sink" soil resource redistribution, leading to highly patchy ground vegetation, and consequently to lower fuel connectivity (Hodgkinson, 2002).

The invasion of introduced grasses in most grasslands in the world as happens with *Cenchrus ciliaris*, a grass with a low inflammability threshold and rapid regeneration may modify fire dynamics with negative consequences for *Acacia* communities (Butler & Fairfax, 2003). Fires under such conditions alter both height and abundance of *Acacia* and other species modifying fire regimes, increasing intensity and frequency of new fire events (Clarke et al., 2005).

Rainy seasons alternated with dry seasons are usually the most favorable climatic conditions for the rise of fire risk. If such conditions are combined with high terrain slopes and windy conditions, fire expansion may become catastrophic and can affect importantly local vegetation survival and recovery.

Such conditions are typical in the central highlands of Argentina, where natural herbivores grow in low densities and the reduction of grass biomass depends almost exclusively on livestock. Fire is the main shaper of these environments; when fires are intense, they usually transform original woodlands of *Lithraea molleoides* and *Schinopsis marginata* into *Acacia s.l.* scrubs, where soil, vegetation, and fauna change their dynamics.

Fires occurring in these *Acacia* scrubs are ordinarily of low to moderate intensity, nevertheless, they may spread for kilometers if connectivity is assured. Fire ignites more frequently in dry and hot conditions when the grasses have water contents lesser than 20% (Karlin, unpublished data; based on Behave Plus fire modeling: Andrews, 2014). If the accumulated woody combustible is unimportant within the system, forest fires tend to be of low intensity, depending exclusively on grass biomass.

Figure 4. *Vachellia* spp. scrub in central Argentina affected by the fire, on the left of the image. The accumulated grass biomass can be seen on the right of the image. © M. Karlin.

Such forest fires do not usually exceed a burning depth of 2 cm below ground (Karlin et al., 2014) and heat temperatures of 200°C (Denegri et al., 2014). Strong recruitment is expected when a significant fraction of the seed bank experiences a heat shock without exceeding lethal limits (Wright et al., 2015). Therophytes, hemicryptophytes, and geophytes are usually not affected by fire or may even be activated by it (Karlin et al., 2016).

In this region, secondary forests are defined by *Vachellia caven*, *V. aroma* and, to a lesser extent, by *Senegalia praecox*. The first two species are co-dominants in this type of substitution forests and enhance the development of grasses with a high competition ability, which tend to cover most of the underbrush because *Vachellia* species do not usually surpass a 30% of soil cover, enabling light to reach understory. *Senegalia praecox* as a non-dominant species is surrogated by *S. marginata*, and undergrowth development usually does not depend on it.

Depending on canopy density and livestock grazing, understory grasses may reach biomass amounts up to 5000 kg of dry matter (Karlin et al., 2015). Local experiences show that biomass amounts over 3000 kg are sufficient to trigger low to mid-intensity fire events.

## Physiological and Ecological Response of *Acacia s.l.* to Fire

Several experiences prove that fire can activate *Acacia s.l.* seed bank. In the Sydney region (Australia), *Acacia* species represent some of the earliest seedlings to emerge after the fire. Apparently, fire activates the dormant seed bank and enhance the appearance of *Acacia suaveolens* seedlings in a high proportion during three months after the fire (Auld & Tozer, 1995). Such *Acacia* pulses, if seedlings prosper, might foment the establishment of coetaneous populations. For plant species that are subject to significant levels of seed loss to predators, disease and/or decay, the timing of seeding events in relation to fire is likely to be an important determinant of post-fire recovery (Wright et al., 2015).

In Uganda, *Vachellia sieberiana* also showed increased germination in burned sites after the beginning of rainfall season, from superficial layers of the affected soil (Sabiiti & Wein, 1987).

Wright et al. (2015) studied heat intensity and length over the activation of the *Acacia aptaneura* seed bank. They report that such seeds show maximum germination rates when the heat reaches the 100°C, suggesting a heat threshold of 70°C in which germination in *A. aptaneura* seeds begins to be stimulated. Seed mortality begins being significant at heat temperatures up to 120°C, increasing with heat length.

Fire severity represents a key constraint for *Acacia* resprouting. Depending on fire heat, it may affect deeper soil layers and with it, deeper resprouting buds. Such fire intensity depends on pre-fuel loads and fire frequency. However, for central Australia *Acacias* growing in spinifex grasslands, buds depth is insufficient for plants to escape from fire and to resprout; they also need sufficient carbohydrate reserves to recover from a fire in summer fire events (Wright & Clarke, 2007).

Resprouting ability also depends on fire frequency and plant age. If fire intervals are short, some *Acacia* species may respond to the disturb producing shallower root buds in secondary shoots because of the previous fire event (Wright & Clarke, 2007). However, resprout ability can decay when plants switch from juvenile to adult phases (Nano et al., 2012).

Seemingly, Australian *Acacia* species with a resprouting ability after fire are poor seed recruiters, and vice versa (Nano et al., 2012). When plants have the ability to recruit, fruits can be more or less fleshy, depending on the species. Fleshy fruit producers are better related to *Acacia* shrublands where there is greater availability of bird perches and are usually produced by non-resprouter species. Seed dissemination may depend on how appetizing the fruits are for the birds; this suggests that some *Acacia* species are linked to specific wildlife (See *Seed Dispersal by Animals* subsection in this chapter).

*Vachellia* species in central Argentina co-habit with native grasses in the context of cyclic fire events. This genus seemingly behaves similarly than Australian *Acacias*. *V. aroma* has the ability to resprout from epicormic buds in the crown and *V. caven* does so by aerial buds that escaped from the effect of the fire, often above the 2 m high (Karlin et al., 2016). Apparently, *V. caven* has also the ability to produce high seed pools that may persist dormant in the soil until some stimuli like fire trigger germination.

In African savannas *Acacia s.l.* species also have resprouting ability after the fire, proportionally with plant height. If the fire affects the canopy of *Senegalia mellifera*, such species might trigger resprouting mechanisms from below ground tissues. Resprouting promotes multi-stemmed shrubs encroaching into open savannas and strongly competing with other species. Curiously, if plants are taller, they have lesser relative regrowth per unit of height after a fire in spite of their having more belowground resources than small plants. Such relation might be due to anatomical (growth constriction of apical meristems) or physiological (lesser photosynthate allocation) mechanisms, not by ecological causes (Meyer et al., 2005).

## Have Acacieae Coevolved with Fire Worldwide?

In the sight of ecological evidence and coincidences among *Acacia s.l.* communities, it seems true, at least for some species and habitats. *Vachellia* species in the central highlands of Argentina, *Vachellia* and *Senegalia* in South Africa savannas and some *Acacia* species in Australian grasslands have positive feedbacks with fire events of low frequency. Nevertheless, studies suggest that fire events should not be highly intense or frequent for *Acacias* to respond with resprouting or seedling recruiting mechanisms.

Nano et al. (2012) suggest that positive feedbacks of fire with *Acacia* communities may apply for semiarid grasslands, but may affect severely the performance of *Acacia* communities from arid shrublands. *Acacia aneura* communities in arid shrublands, for instance, have better performance in seed recruiting when trees are adults and have grown unaffected by disturbances like fire events. Seedling recruiting may be critical for patch maintenance and expansion within arid areas. The development of "fertile islands" represents the prime cause for *Acacia* communities to endure in time and for the future seedling establishment. Fertile island function may be severely compromised by the interaction of fire and extreme weather events, leading to excessive run-off and the creation of "leaky systems" (Hodgkinson, 2002). Anthropized grassland ecosystems have altered fire dynamics in the last centuries because of the evolution of the grazing frequency and intensity and the use of fire as a tool for improving forage. Such phenomena are altering the behavior of biotic communities and may affect the survival of *Acacias*. It may be true that *Acacia s.l.* have co-evolved with grasses and fire events, but it is also true that climate change and land use changes can alter fire dynamics, increasing intensity, and frequency. *Acacias* are in essence intolerant of repeated fire due to slow stem growth/regrowth rates (Nano et al., 2012).

## Can Acacieae Promote Fire-Prone Environments?

The invasion of *Acacia* species (e.g., *A. saligna*, *A. mearnsii*, *A. longifolia*, *A. dealbata*) in several areas of the globe has created fire-prone vegetation features that altered the fire intensity and frequency respect the original ecosystems. Such invasive species can create dense thickets, increasing combustible biomass. Then promoted fire events tend to stimulate germination of *Acacia* seeds in the seed bank, depleting native vegetation (Le Maitre et al., 2011).

In South African fynbos the increase of fire events frequency has conducted to the expanse of *Acacia saligna*, *A. mearnsii* or *A. longifolia* since fire promotes resprouting and seedling recruitment in these species. After fire, *Acacia* species outnumber native species seedlings (Holmes, 2001) or may have faster growth respect native species (Le Maitre et al., 2011) and outcompetes them when grown together.

Originally, in Chilean forests, fire events were rare or less frequent than in the actually invaded ecosystems. Fire reduces the native species cover and reduces the viability of native seeds; it also stimulates epicormic sprouting of *Acacia dealbata*, which tends to increase its cover, reducing richness and abundance of natives (Le Maitre et al., 2011).

## *ACACIA* INVASIVENESS

Plants may show unique dynamics depending on the ecological conditions where they are developing. When plants are introduced into exotic environments, they may or not adapt to new conditions. Most aliens persist temporarily for a short time without a human-assisted input of propagules; these are defined as *casual*. Some aliens may adapt to new conditions, but they do not necessarily spread out; these are defined as *naturalized*. If aliens adopt the ability to spread and suppress natives, they become *invasive* (Rejmánek et al., 2013).

Native plants may also have the ability to spread over their own environments if some disturbing, especially from human cause, modify its steady-state ecosystem reducing the influence of original dominating species. In this case, it may also be possible to implement the concept of invasiveness when dominated species become the dominating in their own niche.

The impacts of invasive plants include disruption to ecological processes by accelerating biomass accumulation, reduced light penetration, increased nitrification, changed fire intensity and frequency, altered geomorphological processes, hybridization with congeners, which can lead to declines in species richness and abundance (Correia, 2012).

The legume subfamily Mimosoideae has a significant percentage of invasive species, and the Australian *Acacia* species have the highest rate of all the legumes. No specific taxonomic characters give the *Acacia* species characteristics of invasiveness; nevertheless, the invasion of *Acacias* as alien species occupying exotic niches has conducted to severe ecological modifications in host ecosystems.

To understand *Acacias* invasiveness, Miller & Burd (2014) propose evaluating the following hypotheses; each one may throw some light to the *Acacias* performance over invaded ecosystems, but also to their evolutionary history. At this point such hypotheses are analyzed (though, not necessarily in the order the authors originally put them).

> *H1: A species will not be able to invade an area that has abiotic (and may add "biotic") conditions that are outside its physiological tolerance levels*
>
> *H3: An invasive species will not be able to replace a native species if they occupy the same niche and are in all other ways equal*

During the adaptive evolution in arid environments, *Acacia s.l.* lineages evolved different structures and mechanisms. Presumably the distinct characteristics of the different lineages provided them the adaptive advantages over the competition. Environmental shifts from closed to open landscapes in Africa and The Americas supported the expansion of grasslands, giving rise to large herbivores (differently than happened in Australia), but also to plants with anti-herbivore adaptations (Miller & Burd, 2014).

It is possible that *Acacia s.l.*, along its radiation and differentiation period, did not invade a saturated community during the major diversification events. Instead, it may have outcompeted other taxa through the development of phytotoxic (allelopathic) compounds, anti-herbivory and water-saving adaptations to come to dominate these slowly evolving landscapes (Lorenzo et al., 2010; Miller & Burd, 2014). It might also co-evolved with local species like the expanding grasses during the Mio-Pliocene, along with the progressive cooling of the atmosphere. With the grasses, there become more frequent fire events that might have forced *Acacias* to adapt into the expanding savannas.

In Australia, evolution occurred differently. Most of the *Acacia s.s.*, in a different evolving environment, did not require "spiny" structures due to the lack of large herbivores. However, the evolution of phyllodes and a thick scleromorphic outer layer for preventing water loss in a xeric environment could be an adaption to success in arid conditions.

Not always invasive species produce negative effects over the host environment. In some cases, exotic Acacieae proved to be beneficial, as described recently for *Vachellia caven* in Central Chile, where it nurses endemic sclerophyllous trees along a successional pathway from silvopastoral savanna to forest (Root-Bernstein et al., 2017).

*H2: The extent of an invasion is negatively correlated to species diversity of functional guild competitors in the invaded environment*

*H8: Species can more easily invade highly disturbed areas; this disturbance can be man-made or natural*

Such hypotheses may be true with Acacieae populations in central Argentina. As seen in the "Plant-Plant Relationships" section in this chapter, most *Vachellia* and *Senegalia* species grow understory of steady-states forests. When original dominant populations are strong within the ecosystem some *Acacia s.l.* species grow as co-dominant or as companion species due possibly to light regulation by the dominants. When dominants are eliminated or reduced (generally, because they have economic importance) disturbance lead to *Acacia* encroachment into secondary forests, becoming the new dominant population and they may even be able to transform their niches (Karlin et al., 2013; Karlin et al., 2015), e.g., by enriching the soil with atmospheric-fixed nitrogen or by increasing litter on the soil. The increase of fire frequencies due to climate change or reductions over livestock loads also favors fire tolerant Acacieae in spite of the usually more fire-sensitive climax-dominant populations (e.g., *Aspidosperma quebracho-blanco* or *Schinopsis marginata*).

Among alien species and considering the extension, aggressiveness, and impact on host ecosystems, Australian *Acacias* are considered the most problematic and widespread invasive plant species worldwide due to their massive introduction into exotic niches for human use (fodder, dune stabilization, ornamentation, etc.) (Lorenzo & Rodriguez-Echeverría, 2015).

The level of invasiveness of many Australian *Acacias* over exotic niches seems to have more to do with human-mediated events and invasibility (susceptibility of the environment to be invaded) of the host environment than with biological features of the species. The widespread use of Mimosoid legumes as forage species and in various types of forestry and agroforestry programs worldwide has radically enhanced their invasiveness potential.

However, invasive *Acacias* have proven to be transformers of the host niches, especially altering soil properties. They produce large amounts of litter and have the ability to produce high amounts of root biomass, altering natural geochemical cycles. When invasive *Acacias* encroach, evapotranspiration can be highly increased reducing the available water and limiting native species growth (Lorenzo & Rodríguez-Echeverría, 2015). Such characteristics gradually displace native (and even supposedly highly resilient) organisms reducing their natural ecological bias.

*H4: A species will not be able to invade an area that harbors pathogens or predators that it has not encountered before.*

*H5: A species will not be able to invade an area if its co-evolutionary species is/are not present in the area.*

After the invasion of *Acacias* into South African Fynbos different Australian parasitoid wasps, *Trichilogaster* spp. were introduced as a biological control. *Trichilogaster* spp. induce galls on the inflorescences of

*Acacia floribunda*, *A. longifolia*, *A. stricta* (*T. acaciaelongifoliae*), and *A. pycnantha* (*T. signiventris*), reducing drastically their reproductive potential in a flowering season (Dennill, 1990; Hoffmann et al., 2002; Marchante et al., 2011). Their introduction had a significant degree of success; however, when the experience was exported to Portugal, it showed some unexpected alternative hosts like *Vitis vinifera* (Marchante et al. 2011).

The ability to fixating nitrogen can modify soil characteristics and with it, the ability of native species to compete with the invasive *Acacias*. As happens with the invasion of the Australian *Acacia longifolia* in Portugal, they can carry with them exotic rhizobia symbionts and replace native rhizobia. Due that exotic rhizobia are less effective in promoting the growth of native legumes, *Acacias* become more effective as invasive species (Rodríguez-Echevarría, 2010).

Invader *Acacias* have also proved to be very good nutrient uptakers when they are associated with mycorrhizal fungi (Lorenzo & Rodríguez-Echeverría, 2015).

Regarding pollination, most Acacieae are protogynous, female flowering precedes male flowering, and consequently, they need cross-pollination by visitors. Acacieae inflorescences are accessible to a broad variety of visitors, and they have a wide flowering period by showing stepped blooming throughout the year and pollen release is gradual. All these characteristics imply an enhanced capacity to invade new locations by extending the number of visits (Stone et al., 2003; Lorenzo et al., 2010). Widespread aggressive pollinators like the European bee can be a dominant vector for Acacieae.

Acacieae seeds are highly dependent on endozoochory for dormancy elimination and spreading. The case relating *Vachellia farnesiana* and the European rabbit has been described in the subsection *Seed Dispersal by Animals*, but other examples can be found. There are some hypotheses asserting that *Vachellia caven* invaded the matorral sclerophyllous forest of central Chile because of the dispersal by transhumance camelids from Argentina before the period of colonies (Ovalle et al., 1990), but European rabbit has also contributed to preventing the re-colonization of clearings by the matorral on the dry slopes (Novillo & Ojeda, 2008).

Ants also may become important vectors for the dissemination of *Acacia* seeds. In the South African Fynbos, small mammals consume seeds, but seeds may be removed by local ants into their nests before mammal predation. These nests become seed reserves that can be activated after fire events (Le Maitre et al., 2011).

*H6: Species that occur at low population densities in their native range will not be invasive.*

There is plenty of evidence that species, which are not invasive nor even dominant in their own original communities, can become invasive in exotic niches. Without competition, abiotic constraints or biological controls alien *Acacias* have no population restrains to spread-out and can displace native organisms.

Such species may not behave as invaders on the first stages of introduction, but some disturbance can alter ecological conditions increasing invasibility. Fire events in the South African Fynbos "reset" the ecosystem, activating the seed-bank. Natives germinate as well as alien *Acacias*; *Acacia* saplings quickly outgrow the fynbos, and these can form dense stands excluding shorter fynbos and no longer provides shelter for small mammals. *Acacia* thickets increase their seed density in the soil and their population expands. In Portuguese coastal dunes or Chilean forests, invasiveness dynamics are remarkably similar (Le Maitre et al., 2011).

The Chilean espinal, with dense populations of *Vachellia caven*, is recognized as a purely anthropogenic feature growing between the 32° to 37°S. This species grows scattered into the nearby Argentinian region of Monte and Puna, from where supposedly the species was introduced to Chile (Ovalle et al., 1990).

*H7: A species will not be able to invade an area if it has a lower use efficiency of its limiting resources than a native species that occupies the same location.*

Water might be one of the most critical resources for which plants compete. The invasion over riparian areas depends more on the velocity of water extraction and evapotranspiration volumes than in the water use efficiency. Ordinarily, adapted native species tend to consume less water than exotic species. Riparian stands of *Acacia mearnsii* in South Africa use more water than native adapted species. However, dense stands of the *Acacias* growing in azonal lands tend to be more or equally efficient in water use than fynbos (Dye & Jarmain, 2004). *A. mearnsii* showed to be a profligate user of water where it is available in large amounts, but more drought-resistant than some native species; therefore, it seems this species has an important competition bias over the natives (Crous et al., 2012).

Generalist species like *Acacia s.l.* would experence no problems adapting to new conditions. Since symbionts (such as rhizobia) or insects (such as pollinators) cooperate unspecifically with *Acacias*, native biological organisms in host niches may perform as effectively as those from native habitats. The spread of pollinator like the European bee (*Apis mellifera*) worldwide must have interacted positively with introduced *Acacias* in exotic ecosystems. Co-existence of multiple *Acacia s.l.* species in plant communities does not appear to be limited by pollinator availability, but when species are sympatric, they may have to adapt to a shared pollinator pool (Miller & Burd, 2014).

*H9: Species from older lineages are more vulnerable to being replaced by invasive species that occupy a similar niche.*

Several authors (Ross, 1981; Miller & Burd, 2014) suggest that the Acacieae differentiation occurred first in *Senegalia*, then in *Vachellia*, and lastly in *Acacia*. According to this hypothesis, the introduction of Australian *Acacia* in The Americas or Africa might prove more successful than backward. However, there are no records regarding competition among different genera within the Acacieae tribe.

However, some cases of sympatry among native species have been found. Among four *Acacieae* species (*Vachellia drepanolobium*, *V. nilotica*, *V. zanzibarica* and *Senegalia senegal*) a daily temporal partitioning of pollinators is cited (Stone et al., 1996). Pollen and nectar release occur differently, with maximum rates of visitation at 6:00, 10:00, 12:00, and

14:00 for *V. drepanolobium*, *V. zanzibarica*, *V. nilotica*, and *S. senegal*, respectively. This mechanism reduces interspecific competition among Acacieae for pollinators. Nevertheless, *S. senegal* has a differential competitive ability respect *Vachellia* spp. which is the production of critical loads of nectar favoring the visitation of honey bees. If nectar production favors competition for *S. senegal*, this fact would prove wrong the present hypothesis.

*H10: A species will only be able to invade an area if it has a life-history strategy which is more r-selected than that of the species which already is occupying the niche.*

The arid habitats into which *Acacia s.l.* radiated requires competitive abilities to cope with the scarcity of resources, especially water. Such environments likely contained originally autochthonous species with K-selected life histories. Thus, *Acacia s.l.* may have been relatively weedier than many of its native competitors (Miller & Burd, 2014).

The ability of most Acacieae species to cover the soil rapidly after fire events suggests a substantial competitive advantage. In the **Fire Ecology** section, it was mentioned that some *Acacia s.l.* species break seed dormancy after low-intensity events (therophyte species), or that stimulates epicormic sprouting in other (hemicryptophytes or phanerophytes –Figure 5– species), growing faster than most species co-habiting them (Karlin et al., 2016), and increasing competitive ability.

*H11: There are no rules concerning whether a species is invasive or not; it all happens by chance.*

As conclusion, it is possible to assert that the patterns of invasion observed in the field at one site may be difficult to extrapolate to other sites. However, after all the evidence shown in this chapter, there are some rules that operate during the invasion.

*Acacias* are generalist species that may adapt to very dissimilar ecological conditions. They have an exceptional ability to colonize rapidly empty niches through fast-growing strategies and through a wide variety of vectors. They are usually transformer species that can modify the environment by nitrogen fixation or litter accumulation, altering soil dynamics. They may relate to different biological organisms, like bacteria, mycorrhizal fungi, ants, bees, birds, ungulates, among others. They may activate collaborative or symbiotic mechanisms than help Acacieae to increase competitive abilities over neighboring plants.

Of course, each species within the Acacieae tribe may adapt according to local ecological characteristics differently, and they even may differ in its behavior when growing in different environments or ecosystems.

The evidences shown in this chapter may help to integrate several aspects of Acacieae ecology and may even help understand the behavior of other related species.

Figure 5. Epicormic sprouting of *Vachellia caven*, two weeks after a fire event, Sierras de Córdoba (Argentina). © M. Karlin.

# REFERENCES

Al-Mefarrej, H., (2012). Diversity and frequency of *Acacia* spp. in three regions in the Kingdom of Saudi Arabia. *Afr. J. Biotech.*, 11(52), 11420-11430.

Andrews, P. L., (2014). Current status and future needs of the BehavePlus Fire Modeling System. *Int. J. Wildland Fire*, 23(1), 21-33.

Aronson, J. (1992). Evolutionary biology of *Acacia caven* (Leguminosae, Mimosoideae): infraspecific variation in fruit and seed characters. *Ann. Missouri Bot. Gard.*, 79(4), 958-968.

Auld, T. D., Tozer, M., (1995). Patterns in emergence of *Acacia* and *Grevillea* seedlings after fire. *Proc. Linn. Soc. NSW* 1(15), 5-15.

Avis, A. M., (1989). A review of coastal dune stabilization in the Cape Province of South Africa. *Land. Urb. Plan.*, 18(1), 55-68.

Barnes, M. E., (2001). Effects of large herbivores and fire on the regeneration of *Acacia erioloba* woodlands in Chobe National Park, Botswana. *Afr. J. Ecol.*, 39(4), 340-350.

Bond, W. J., Midgley, G. F., & Woodward, F. I., (2003). The importance of low atmospheric $CO_2$ and fire in promoting the spread of grasslands and savannas. *Glob. Change Biol.*, 9(7), 973-982.

Bouchenak-Khelladi, Y., Maurin, O., Hurter, J., Van der Bank, M., (2010). The evolutionary history and biogeography of Mimosoideae (Leguminosae): an emphasis on African *Acacias*. *Mol. Phylogenet. Evol.*, 57(2), 495-508.

Brandes, A. F., Barros, C. F., (2007). Growth rings in lianas of Leguminosae family from Atlantic forest in Rio de Janeiro, Brazil. In *Anatomia do lenho e dendrocronologia de lianas da família leguminosae ocorrentes na mata atlântica*, Master's Degree thesis, 42-67. [In *Anatomy of the wood and dendrochronology of lianas of the legume family and occurring in the Atlantic Forest*, Master's Degree thesis, 42-67].

Brooks, R., Owen-Smith, N., (1994). Plant defenses against mammalian herbivores: are juvenile *Acacia* more heavily defended than mature trees? *Bothalia*, 24(2), 211-215.

Brown, W. L., (1960). Ants, *Acacias* and browsing mammals. *Ecology*, 41(3), 587-592.

Bucher, E. H., (1987). Herbivory in arid and semi-arid regions of Argentina. *Rev. Chil. Hist. Nat.*, 60, 265-273.

Bui, E. N., González-Orozco, C. E., Miller, J. T., (2014). *Acacia*, climate, and geochemistry in Australia. *Plant Soil*, 381(1-2), 161-175.

Butler, D. W., Fairfax, R. J., (2003). Buffel Grass and fire in a Gidgee and Brigalow woodland: a case study from central Queensland. *Ecol. Manage. Rest,* 4(2), 120–125.

Castellanos-Barliza, J., León Peláez, J. D., (2011). Descomposición de hojarasca y liberación de nutrientes en plantaciones de *Acacia mangium* (Mimosaceae) establecidas en suelos degradados de Colombia. *Rev. Biol. Trop.*, 59(1), 113-128. [Leaf litter decomposition and nutrient release in plantations of *Acacia mangium* (Mimosaceae) established in degraded soils of Colombia. *Tropical Biology Journal*, 59(1), 113-128].

Chou, C. B., Hedin, L. O., Pacala, S. W., (2018). Functional groups, species and light interact with nutrient limitation during tropical rainforest sapling bottleneck. *J. Ecol.*, 106(1), 157-167.

Clarke, P. J., Latz, P. K., Albrecht, D. E., (2005). Long-term changes in semi-arid vegetation: invasion of an exotic perennial grass has larger effects than rainfall variability. *J. Veg. Sci.* 16(2), 237–248.

Correia, M. C. L., (2012). *Reproductive biology of Australian Acacias in Portugal*. Dto. de Ciências da Vida, Fac. De Ciências e Tecnologia, Univ de Coimbra, Portugal. Master's Degree Thesis.

Crous, C. J., Jacobs, S. M., Esler, K. J., (2012). Drought-tolerance of an invasive alien tree, *Acacia mearnsii* and two native competitors in fynbos riparian ecotones. *Biol. Inv.*, 14(3), 619-631.

Davidson, D. W., Morton, S. R., (1984). Dispersal adaptations of some *Acacia* species in the Australian arid zone. *Ecology*, 65(4), 1038-1051.

Dean, C., Fitzgerald, N. B., Wardell-Johnson, G. W., (2011). Pre-logging carbon accounts in old-growth forests, via allometry: An example of mixed-forest in Tasmania, Australia. *Plant Biosystems-An International Journal Dealing with all Aspects of Plant Biology*, 146(1), 223-236.

Dean, W. R. J., Milton, S. J., Jeltsch, F., (1999). Large trees, fertile islands, and birds in arid savanna. *J. Arid Environ.*, 41(1), 61-78.

Denegri, A., Toranzo, L., Rubenacker, A., Campitelli, P., Karlin, M., (2014). Efecto de los incendios forestales sobre las propiedades del suelo. *NEXO Agrop.*, 2(1-2), 10-14. [Effect of forest fires on soil properties. *NEXO Agrop.*, 2(1-2), 10-14].

Dennill, G. B., (1990). The contribution of a successful biocontrol project to the theory of agent selection in weed biocontrol—the gall wasp *Trichilogaster acaciaelongifoliae* and the weed *Acacia longifolia*. *Agr. Ecosyst. Environ.*, 31(2), 147-154.

Dohn, J., Dembélé, F., Karembé, M., Moustakas, A., Amévor, K. A., Hanan, N. P., (2013). Tree effects on grass growth in savannas: competition, facilitation and the stress-gradient hypothesis. *J. Ecol.*, 101(1), 202-209.

Dye, P., Jarmain, C., (2004). Water use by black wattle (*Acacia mearnsii*): implications for the link between removal of invading trees and catchment streamflow response: working for water. *S. Afr. J. Sci.*, 100(1-2), 40-44.

Dynes, R. A., Schlink, A. C., (2002). Livestock potential of Australian species of *Acacia*. *Cons. Sci. W. Aust.*, 4(3), 117-124.

El-Bana, M. I., Al-Mathnani, A., (2009). Vegetation-soil relationships in the Wadi Al-Hayat area of the Libyan Sahara. *Aust. J. Basic Appl. Sci.*, 3(2), 740-747.

Ferreira Barros, M. J., Pires Morim, M., (2014). *Senegalia* (Leguminosae, Mimosoideae) from the Atlantic Domain, Brazil. *Syst. Bot.*, 39(2), 452-477.

Gaddis, K. D., (2014). *The population biology of dispersal and gene flow in the desert shrub Acacia (Senegalia) greggii A. Gray in the Mojave National Preserve.* University of California, Los Angeles, PhD Degree Thesis.

Gilbert, J. M., (1959). Forest succession in the Florentine Valley, Tasmania. *Pap. Proc. Royal Soc. Tasmania*, 93, 129-152.

Gowda, J. H., (1997). Physical and chemical response of juvenile *Acacia tortilis* trees to browsing. Experimental evidence. *Func. Ecol.*, 11(1), 106-111.

Gowda, J. H., Albrectsen, B. R., Ball, J. P., Sjöberg, M., Palo, R. T., (2003). Spines as a mechanical defense: the effects of fertilizer treatment on juvenile *Acacia tortilis* plants. *Acta Oecol.*, 24(1), 1-4.

Guinet, P., Vassal, J., (1978). Hypotheses on the differentiation of the major groups in the genus *Acacia* (Leguminosae). *Kew Bull.*, 509-527.

Helman, D., Leu, S., Mussery, A., (2017). Contrasting effects of two *Acacia* species on understorey growth in a drylands environment: Interplay of canopy shading and litter interference. *J. Veg. Sci.*, 28(6), 1140-1150.

Hodgkinson, K. C., (2002). Fire in *Acacia* wooded landscapes: effects on functional processes and biological diversity. In Bradstock, R. B., Williams, J. E., Gill, A. M. (Eds.) *Flammable Australia: Fire regimes and the biodiversity of a continent*. Cambridge University Press, Cambridge, UK: 259–280.

Hoffmann, J. H., Impson, F. A. C., Moran, V. C., Donnelly, D., (2002). Biological control of invasive golden wattle trees (*Acacia pycnantha*) by a gall wasp, *Trichilogaster* sp. (Hymenoptera: Pteromalidae), in South Africa. *Biol. Control*, 25(1), 64-73.

Holmes, P.M., (2001). A comparison of the impacts of winter versus summer burning of slash fuel in alien-invaded fynbos areas in the Western Cape. *S. Afr. Forest. J.*, 192, 41–50.

Karban, R., Agrawal, A. A., Thaler, J. S., Adler, L. S., (1999). Induced plant responses and information content about risk of herbivory. *Trends Ecol. Evol.*, 14(11), 443-447.

Karlin, M, Arnulphi, S, Alday, A., Bernasconi, J., Accietto, R., (2016). Revegetación post-incendio en matorrales de *Acacia* spp. en las Sierras de Córdoba, Argentina Central. *Oecol. Aust.*, 20(4), 464-476. [Post-fire revegetation in *Acacia* spp. shrublands in Sierras of Córdoba, Central Argentina. *Oecol. Aust.*, 20(4), 464-476].

Karlin, M., Arnulphi, S., Alday, A., Bernasconi, J., Accietto, R., Denegri, A., Rubenacker, A., Toranzo, L., Campitelli, P., (2014). Efecto de la aplicación de compost sobre la revegetación natural de suelos afectados por incendios. *NEXO Agrop.*, 2(1-2), 6-9. [Effect of the application of compost on the natural revegetation of fire-affected soils. *NEXO Agrop.*, 2(1-2), 6-9].

Karlin, M., Bernasconi, J., Schneider, C., Rufini, S., Accietto, R., Arnulphi, S., (2015). Aprovechamiento de la potencialidad silvopastoril como alternativa para el control de incendios en la Reserva Natural Militar La Calera, Córdoba (Argentina). *III Congreso Nacional de Sistemas Silvopastoriles* y *VIII Congreso Internacional de Sistemas Agro-*

*forestales*. Iguazú, Misiones: 636-640. [The use of silvopastoral potential as an alternative for fire control in the Military Natural Reserve of La Calera, Córdoba (Argentina). *III National Congress of Silvopastoral Systems and VIII International Congress of Agroforestry Systems*. Iguazú, Misiones: 636-640].

Karlin, M., Galán, R., Contreras, A., Zapata, R., Coirini, R., Ruiz Posse, E., (2013). Exergetic model of secondary successions for plant communities in Arid Chaco (Argentina). *ISRN Biodivers.*, 2013.

Kaur, B., Gupta, S. R., Singh, G., (2000). Soil carbon, microbial activity and nitrogen availability in agroforestry systems on moderately alkaline soils in northern India. *Appl. Soil Ecol.*, 15(3), 283-294.

Kos, M., Poschlod, P., (2007). Seeds use temperature cues to ensure germination under nurse-plant shade in xeric Kalahari savannah. *Ann. Bot.*, 99(4), 667-675.

Kriticos, D. J., Sutherst, R. W., Brown, J. R., Adkins, S. W., Maywald, G. F., (2003). Climate change and the potential distribution of an invasive alien plant: *Acacia nilotica* ssp. *indica* in Australia. *J. Appl. Ecol.*, 40(1), 111-124.

Kürschner, W. M., Kvaček, Z., Dilcher, D. L., (2008). The impact of Miocene atmospheric carbon dioxide fluctuations on climate and the evolution of terrestrial ecosystems. *Proc. Nat. Acad. Sci.*, 105(2), 449-453.

Kuyper, T. W., de Goede, R. G., (2013). Interaction between higher plants and soil-dwelling organisms. In van der Maarel, E., Franklin, J. (Eds.). *Vegetation Ecology*, Wiley-Blackwell, UK: 260-284.

Le Maitre, D. C., Gaertner, M., Marchante, E., Ens, E. J., Holmes, P. M., Pauchard, A., O'Farrel, P. J., Rogers, A. M., Banchard, M., Blignaut, J., Richardson, D. M., (2011). Impacts of invasive Australian *Acacias*: implications for management and restoration. *Divers. Distrib.*, 17(5), 1015-1029.

Linstädter, A., Bora, Z., Tolera, A., Angassa, A., (2016). Are trees of intermediate density more facilitative? Canopy effects of four East African legume trees. *Appl. Veg. Sci.*, 19(2), 291-303.

Lorenzo, P., González, L., Reigosa, M. J., (2010). The genus *Acacia* as invader: the characteristic case of *Acacia dealbata* Link in Europe. *Ann. For. Sci.*, 67(1), 101.

Lorenzo, P., Rodríguez-Echeverría, S., (2015). Cambios provocados en el suelo por la invasión de *Acacias* australianas. *Rev. Ecosist.*, 24(1), 59-66. [Changes caused in the soil by the invasion of Australian *Acacias*. *Ecosystems Journal*, 24(1), 59-66].

Ludwig, F., Dawson, T. E., de Kroon, H., Berendse, F., Prins, H. H., (2003). Hydraulic lift in *Acacia tortilis* trees on an East African savanna. *Oecol.*, 134(3), 293-300.

Ludwig, F., Dawson, T. E., Prins, H. H. T., Berendse, F., Kroon, H. (2004). Below-ground competition between trees and grasses may overwhelm the facilitative effects of hydraulic lift. *Ecol. Lett.*, 7(8), 623-631.

Marchante, H., Freitas, H., Hoffmann, J. H., (2011). Assessing the suitability and safety of a well-known bud-galling wasp, *Trichilogaster acaciaelongifoliae*, for biological control of *Acacia longifolia* in Portugal. *Biol. Control*, 56(2), 193-201.

Maslin, B. R., (2008). Generic and subgeneric names in *Acacia* following retypification of the genus. *Muelleria*, 26(1), 7-9.

Meyer, K. M., Ward, D., Moustakas, A., Wiegand, K., (2005). Big is not better: small *Acacia mellifera* shrubs are more vital after fire. *Afr. J. Ecol.*, 43(2), 131-136.

Mildenhall, D. C., (1972). Fossil pollen of *Acacia* type from New Zealand. *New Zealand J. Bot.*, 10(3), 485-494.

Miller, J. T., Burd, M., (2014). Australia's *Acacia*: Unrecognized convergent evolution. In Prins, H. H. T., Gordon, I. J. (Ed). *Invasion biology and ecosystem theory; Insights from a continent in transformation.* Cambridge University Press, Cambridge and New York: 23-38.

Moore, G., Smith, G. F., Figueiredo, E., Demissew, S., Lewis, G., Schrire, B., Rico, L., van Wyk, A. E., Luckow, M., Kiesling, R., Sousa, M. S., (2011). The *Acacia* controversy resulting from minority rule at the Vienna Nomenclature Section: Much more than arcane arguments and complex technicalities. *Taxon*, 60(3), 852-857.

Nano, C. E., Clarke, P., Pavey, C. R., (2012). Fire regimes in arid hummock grasslands and *Acacia* shrublands. In Bradstock, R. A., Gill, A. M., Williams, R. J. (Eds.). *Flammable Australia: fire regimes, biodiversity, and ecosystems in a changing world*. CSIRO Publishing, Australia: 195-214.

Novillo, A., Ojeda, R. A., (2008). The exotic mammals of Argentina. *Biol. Inv.*, 10(8), 1333.

Obeid, M., El Din, A. S., (1971). Ecological studies of the vegetation of the Sudan. III. The effect of simulated rainfall distribution at different isohyets on the regeneration of *Acacia senegal* (L.) Willd. on clay and sandy soils. *J. Appl. Ecol.*, 8(1), 203-209.

Odee, D. W., Telford, A., Wilson, J., Gaye, A., Cavers, S., (2012). Plio-Pleistocene history and phylogeography of *Acacia senegal* in dry woodlands and savannahs of sub-Saharan tropical Africa: evidence of early colonisation and recent range expansion. *Heredity*, 109(6), 372-382.

Or, K., Ward, D., (2003). Three-way interactions between *Acacia*, large mammalian herbivores, and bruchid beetles-a review. *Afr. J. Ecol.*, 41(3), 257-265.

Ovalle, C., Aronson, J., Del Pozo, A., Avendano, J., (1990). The espinal: agroforestry systems of the Mediterranean-type climate region of Chile. *Agroforestry Syst.*, 10(3), 213-239.

Palmer, T. M., (2003). Spatial habitat heterogeneity influences competition and coexistence in an African *Acacia* ant guild. *Ecology*, 84(11), 2843-2855.

Palmer, T. M., Riginos, C., Damiani, R. E., Morgan, N., Lemboi, J. S., Lengingiro, J., Ruiz-Guajardo, J. C., Pringle, R. M., (2017). Influence of neighboring plants on the dynamics of an ant–*Acacia* protection mutualism. *Ecology*, 98(12), 3034-3043.

Pascual, M. S., Lugo, S. F., Cigala, A. N., (2009). Interaction between two exotic invading species: endozoochory of *Acacia farnesiana* seeds by the European rabbit (*Oryctolagus cuniculus*). *Open Forest Sci. J.*, 2, 000-000.

Pedraza Olivera, R. M., (2008). Metabolitos secundarios no fenólicos en el follaje de árboles y arbustos. Efecto en la fisiología digestiva de rumiantes. *Rev. Prod. Anim.*, 20(2), 97-102. [Non-phenolic secondary metabolites in the foliage of trees and shrubs. Effect on the digestive physiology of ruminants. *Animal Production Journal*, 20(2), 97-102].

Raine, N. E., Willmer, P., Stone, G. N., (2002). Spatial structuring and floral avoidance behavior prevent ant–pollinator conflict in a Mexican ant-*Acacia*. *Ecology*, 83(11), 3086-3096.

Rejmánek, M., Richardson, D. M., Pyšek, P., (2013). Plant invasions and invasibility of plant communities. In van der Maarel, E., Franklin, J. (Eds.) *Vegetation Ecology*. Wiley-Blackwell, UK: 387-424.

Rivers, M.C., (2017). *Acacia nilotica*. The IUCN Red List of Threatened Species 2017. e.T158379A774377.http://dx.doi.org/10.2305/IUCN.UK.2017-3.RLTS.T158379A77437 7.en.

Rodríguez-Echeverría, S., (2010). Rhizobial hitchhikers from Down Under: invasional meltdown in a plant–bacteria mutualism? *J. Biogeogr.*, 37(8), 1611-1622.

Rohner, C., Ward, D., (1997). Chemical and mechanical defense against herbivory in two sympatric species of desert *Acacia*. *J. Veg. Sci.*, 8(5), 717-726.

Rohner, C., Ward, D., (1999). Large mammalian herbivores and the conservation of arid *Acacia* stands in the Middle East. *Cons. Biol.*, 13(5), 1162-1171.

Root-Bernstein, M., Valenzuela, R., Huerta, M., Armesto, J., Jaksic, F., (2017). *Acacia caven* nurses endemic sclerophyllous trees along a successional pathway from silvopastoral savanna to forest. *Ecosphere*, 8(2).

Ross, J. H., (1981). An analysis of the African *Acacia* species: their distribution, possible origins, and relationships. *Bothalia*, 13(3/4), 389-413.

Ross, Z., Burt, J. (2015). Unusual canopy architecture in the umbrella thorn *Acacia*, *Vachellia tortilis* (= *Acacia tortilis*), in the United Arab Emirates. *J. Arid Env.*, 115, 62-65.

Rossi, C. A., De León, M., González, G. L., Pereyra, A. M., (2007). Presencia de metabolitos secundarios en el follaje de diez leñosas de ramoneo en el bosque xerofítico del Chaco Árido Argentino. *Trop. Subtrop. Agroecosyst.*, 7(2), 133-143. [Secondary metabolites presence in ten browse woody plants in the xerophitic woodland in the Argentine Arid Chaco region. *Trop. Subtrop. Agroecosyst.*, 7(2), 133-143].

Sabiiti, E. N., Wein, R. W., (1987). Fire and *Acacia* seeds: a hypothesis of colonization success. *J. Ecol.*, 75(4), 937-946.

Scogings, P. F., Hattas, D., Skarpe, C., Hjältén, J., Dziba, L., Zobolo, A., Rooke, T., (2015). Seasonal variations in nutrients and secondary metabolites in semi-arid savannas depend on year and species. *J. Arid Env.*, 114, 54-61.

Seigler, D. S., Ebinger, J. E., Riggins, C. W., Terra, V., Miller, J. T., (2017). *Parasenegalia* and *Pseudosenegalia* (Fabaceae): New genera of the Mimosoideae. *Novon*, 25(2), 180-205.

Sensenig, R. L., Kimuyu, D. K., Ruiz Guajardo, J. C., Veblen, K. E., Riginos, C., Young, T. P., (2017). Fire disturbance disrupts an *Acacia* ant-plant mutualism in favor of a subordinate ant species. *Ecology*, 98(5), 1455-1464.

Stone, G. N., Raine, N. E., Prescott, M., Willmer, P. G. (2003). Pollination ecology of *Acacias* (Fabaceae, Mimosoideae). *Aust. Syst. Bot.*, 16(1), 103-118.

Stone, G., Willmer, P., Nee, S., (1996). Daily partitioning of pollinators in an African *Acacia* community. *Proc. R. Soc. Lond. B.*, 263(1375), 1389-1393.

Thiele, K. R., Funk, V. A., Iwatsuki, K., Morat, P., Peng, C. I., Raven, P. H., Sarukhan, J., Seberg, O., (2011). The controversy over the retypification of *Acacia* Mill. with an Australian type: a pragmatic view. *Taxon*, 60(1), 194-198.

Toledo-Aceves, T., Swaine, M. D., (2008). Above-and below-ground competition between the liana *Acacia kamerunensis* and tree seedlings in contrasting light environments. *Plant Ecol.*, 196(2), 233-244.

Venier, P., García, C. C., Cabido, M., Funes, G., (2012). Survival and germination of three hard-seeded *Acacia* species after simulated cattle ingestion: The importance of the seed coat structure. *S. Afr. J. Bot.*, 79, 19-24.

Ward, P. S., Branstetter, M. G., (2017). The *Acacia* ants revisited: convergent evolution and biogeographic context in an iconic ant/plant mutualism. *Proc. R. Soc. B.*, 284(1850), 2016-2569.

Whitney, K. D., Stanton, M. L., (2004). Insect seed predators as novel agents of selection on fruit color. *Ecology*, 85(8), 2153-2160.

Wigley, B. J., Bond, W. J., Fritz, H., Coetsee, C., (2015). Mammal browsers and rainfall affect *Acacia* leaf nutrient content, defense, and growth in South African savannas. *Biotropica*, 47(2), 190-200.

Willmer, P. G., Stone, G. N., (1997). How aggressive ant-guards assist seed-set in *Acacia* flowers. *Nature*, 388(6638), 165.

Wright, B. R., Clarke, P. J., (2007). Resprouting responses of *Acacia* shrubs in the Western Desert of Australia–fire severity, interval and season influence survival. *Int. J. Wildland Fire*, 16(3), 317-323.

Wright, B. R., Latz, P. K., Zuur, A. F., (2015). Fire severity mediates seedling recruitment patterns in slender mulga (*Acacia aptaneura*), a fire-sensitive Australian desert shrub with heat-stimulated germination. *Plant Ecol.*, 217(6), 789-800.

Xiong, Y., Xia, H., Li, Z. A., Cai, X. A., Fu, S. (2008). Impacts of litter and understory removal on soil properties in a subtropical *Acacia mangium* plantation in China. *Plant Soil*, 304(1-2), 179-188.

Young, T. P., (1987). Increased thorn length in *Acacia depranolobium*—an induced response to browsing. *Oecol.* 71(3), 436-438.

Young, T. P., Okello, B. D., (1998). Relaxation of an induced defense after exclusion of herbivores: spines on *Acacia drepanolobium*. *Oecol.* 115(4), 508-513.

## Web Pages

*Catalogue of Life*: http://www.catalogueoflife.org. Accessed May 2018
*Encyclopedia of Life*: http://eol.org. Accessed April 2018.

*Global Biodiversity Information Facility*: http://www.gbif.org. Accessed May 2018.

Trópicos.org: http://www.tropicos.org. Accessed May 2018.

*World Wide Wattle*: http://www.worldwidewattle.com. Accessed March 2018.

## BIOGRAPHICAL SKETCHES

### *Marcos Sebastián Karlin*

**Affiliation:** Universidad Nacional de Córdoba, Facultad de Ciencias Agropecuarias, Natural Resources Department, Córdoba, Argentina.

**Education:**
- Agronomic Engeneer. Facultad de Ciencias Agropecuarias. Universidad Nacional de Córdoba, Argentina. 2004.
- Training Program for Biodiversity Information Systems. JICA, Japan. 2012.
- Doctorate in Agricultural Sciences. Facultad de Ciencias Agropecuarias. Universidad Nacional de Córdoba, Argentina. 2013.

**Research and Professional Experience:**
- Technical staff of the Project *Alternativas de sustentabilidad del bosque nativo del Espinal* [Alternatives for the sustainability of native forests of the Espinal], PIARFON, UNC-UNER. 2004.
- Technical staff in the Project *Producción múltiple y diversa bajo bosque en la Provincia del Chaco, Argentina, experiencia Cabeza de Buey y Nueva Pompeya* [Multiple and diverse understory production in the Province of Chaco, Argentina, experience Cabeza de Buey and Nueva Pompeya], Red Agroforestal Chaco-AVINA. 2004-2005.

- Technical staff for the Social Agrarian Program, Catamarca, Argentina. 2006-2007.
- Technical staff for the Project *Manejo sustentable del ecosistema Salinas Grandes del Chaco Árido* [Sustainable Management of the Ecosystem Salinas Grandes of the Arid Chaco], Global Environment Facility, Argentina. 2007-2010.
- Assistant Consultant for the *Manual of Forestry Good Practices and Sustainable Models for the ecoregions of Monte and Espinal*, UNIQUE GmbH. Proyecto de Bosques Nativos y su Sustentabilidad [Native Forests and Sustainability Project], Secretaría de Ambiente y Desarrollo Sustentable de la Nación. 2011-2012.
- Technical staff of the Project *Implementación de actividades sustentables en el área de amortiguamiento del Parque Nacional Talampaya, con las comunidades de La Torre, El Chiflón y Salinas de Busto* [Implementation of sustanable activities in the Talampaya National Park buffer area, with the communities of La Torre, El Chiflón and Salinas de Busto]. Financed and evaluated by the National Parks Administration. 2011-2014.
- Technical staff of the Project *Reevaluación, promoción y manejo de Cercidium praecox (brea) y sus exudados como producto forestal no maderable en el Chaco Árido Argentino* [Reevaluation, promotion and management of *Cercidium praecox* and its exudates as a forestry non-woody product in the Argentinian Arid Chaco]. Programa Nacional de Protección de Bosques Nativos, Secretaría de Ambiente y Desarrollo Social. 2011-2017.
- Academic Coordinator for the Project *El manejo como herramienta para la recuperación de cuencas* [The management as a tool for the recovery of hydrographic basin], financed by AVINA. El Cuenco Equipo Ambiental. 2012 to date.
- Technical staff of the NODO Regional Forestal Monte y Espinal [Monte and Espinal forestry ecoregions NODO]. 2013-2017.
- Director of the Project *Incendios forestales: efectos sobre el suelo y la vegetación nativa* [Forest fires: effects over soil and native vegetation], PROIINDIT. 2013-2015.

- Technical staff of the Project *Uso y manejo no maderero del recurso forestal para la gestión integral de ecosistemas de zonas áridas* [Non-woody use and management of forestry resources for the integral arrangement of ecosystems of arid zones]. Secretaría de Ambiente y Desarrollo Sustentable de la Nación (SAyDS), Proyecto PNUD N° ARG/12/2013. 2013-2016.
- Technical staff of the Project *Calidad de compost para su utilización como enmiendas orgánicas en suelo de uso agrícola* [Compost quality for its use as organic amendments in agricultural soils], SECyT, Universidad Nacional de Córdoba. 2014-2016.
- Post degree Professor of Gestión y Manejo de Zonas Áridas. [Management of Arid Lands]. UNdeC. November 2016.
- Technical staff of the Project *Impacto de las diferentes temperaturas del fuego sobre variables edáficas y de vegetación: diagnóstico y recuperación mediante aplicación de enmiendas orgánicas* [Impact of different fire temperatures on soil and plant variables: diagnosis and recovery with organic amendments], SECyT, Universidad Nacional de Córdoba. 2016-2018.
- Director of the Project *Estudio del componente herbáceo de sistemas ganaderos extensivos con información derivada de sensores remotos* [Study of the herbaceous component in extensive livestock systems with information derived from remote sensors], PROINDIT 2017-2019.

**Professional Appointments:**
- Assistant Professor in the General and Inorganic Chemistry chair, Natural Resources Department, Facultad de Ciencias Agropecuarias, Universidad Nacional de Córdoba. August 2006 to date.
- Associate Professor in Edaphology and Soil Management, Escuela Superior Integral de Lechería (ESIL), Villa María, Córdoba, Argentina. 2006-2011.

- Professor and Tutor in the area of Planning and Management of Hydrographic Basins, Engineering and Rural Mechanization Department, Facultad de Ciencias Agropecuarias, Universidad Nacional de Córdoba, August 2013 to date.
- Co-founder and member of the Asociación Civil El Cuenco-Equipo Ambiental [El Cuenco-Environmental Team, Civil Association]. 2013 to date.
- Professor and Tutor in the area of Management of Natural Resources in Marginal Agrosystems, Natural Resources Department, Facultad de Ciencias Agropecuarias, Universidad Nacional de Córdoba, August 2014 to date.
- Professor in the area of Green Spaces, Engineering and Rural Mechanization Department, Facultad de Ciencias Agropecuarias, Universidad Nacional de Córdoba, August 2016 to date.
- Professor in the Optative Seminary II, Technical School of Gardening and Floriculture, Facultad de Ciencias Agropecuarias, Universidad Nacional de Córdoba. 2016-2017.

**Honors:** Substitute councelor of the Honorable Directive Council, Facultad de Ciencias Agropecuarias-UNC. 2014-2016.

**Publications from the Last 3 Years:**
*Books:*

Carreras, J.; Mazzuferi, V. y M. S. Karlin (eds.). 2016. *El cultivo de garbanzo (Cicer arietinum L.) en Argentina*. Editorial Universidad Nacional de Córdoba. ISBN 978-950-33-1251-3. [*The cultivation of chickpea (Cicer arietinum L.) in Argentina.*]

Karlin, M. S. (ed.); Arnulphi, S. A; Karlin, U. O.; Bernasconi, J. R.; Accietto, R. y A. Cora. 2017. *Plantas del Centro de Argentina*. Ecoval Editorial. ISBN 978-987-4003-14-0. [*Plants of the Center of Argentina.*]

*Book Chapters:*

Bernasconi, J.; Karlin, M.; Accietto, R.; Schneider, C.; Rufini, S. y S. Arnulphi. 2015. Modelos de estados y transiciones: bases para el manejo de la vegetación en la Reserva Natural de la Defensa La Calera, Córdoba, Argentina. In: Martinez Carretero, E. M. & A. Dalmasso (Eds.). *Restauración ecológica en la diagonal árida de la Argentina 2*. Pp.: 3-20. [State models and transitions: bases for the management of vegetation in the La Calera Defense Nature Reserve, Córdoba, Argentina. In: *Ecological restoration in the arid diagonal of Argentina*]

Karlin, M. S. 2016. Manejo de suelo. In: Carreras, J., Mazzuferi, V. & M. Karlin (Eds.). *El cultivo de garbanzo (Cicer arietinum L.) en Argentina*. Editorial Universidad Nacional de Córdoba. Pp.: 57-76. [T*he cultivation of chickpea (Cicer arietinum L.) in Argentina.*]

Karlin, M. S. 2016. Soil-Plant Relationships in the Sabkhat of America. In: Khan, M. A., Boër, B., Özturk, M., Clüsener-Godt, M., Gul, B., & S. W. Breckle (Eds.). *Sabkha Ecosystems Volume V: The Americas*. Volume 48 of the series Tasks for Vegetation Science. Springer. Pp.: 329-347.

Karlin, U. O. & M. Karlin. 2017. Región Humedales valliserranos. 6a. Subregión Ríos y arroyos de los valles intermontanos. In: Benzaquén, L.; Blanco, D., Bo, R., Kandus, P., Lingua, G., Minotti, P. & R. Quintana (Eds.). *Regiones de Humedales de la Argentina*. Fundación para la Conservación y el Uso Sustentable de los Humedales-Wetlands International. Bs. As. ISBN 978-987-29811-6-7. Pp.: 163-172. [Vallliserran Wetlands Region. 6a. Subregion Rivers and streams of the intermontane valleys. In: *Wetlands Regions of Argentina*.]

Karlin, M. & U. O. Karlin. 2017. Región Humedales valliserranos. 6c. Subregión Salinas de la Depresión Central. In: Benzaquén, L.; Blanco, D., Bo, R., Kandus, P., Lingua, G., Minotti, P. & R. Quintana (Eds.). *Regiones de Humedales de la Argentina*. Fundación para la Conservación y el Uso Sustentable de los Humedales-Wetlands International. Bs. As. ISBN 978-987-29811-6-7. Pp.: 185-194. [Vallliserran Wetlands Region. 6c. Subregion Salinas of the Central Depression. In: *Wetlands Regions of Argentina*.]

*Scientific Papers:*

Karlin, M. S. 2016. Ethnoecology, ecosemiosis and integral ecology in Salinas Grandes (Argentina). *Etnobiología* 14(1): 23-38.

Karlin, M; Arnulphi, S; Alday, A.; Bernasconi, J. & R. Accietto. 2016. Revegetación post-incendio en matorrales de *Acacia* spp. en las Sierras de Córdoba, Argentina Central. *Oecologia Australis* 20(4): 464-476. [Post-fire revegetation in thickets of *Acacia* spp. in the Sierras de Córdoba, Central Argentina. *Oecologia Australis*]

Coirini, R.; Karlin, M.; Llaya, G.; Sánchez, S.; Contreras, A. & R. Zapata. 2017. Evaluación de prácticas de desmonte selectivo y clausuras temporales en sistemas degradados del Chaco Árido (Argentina). *Revista de Ciencias Ambientales* (Costa Rica) 51(2): 73-90. [Evaluation of selective clearing practices and temporary closures in degraded systems of the Arid Chaco (Argentina). *Journal of Environmental Sciences*]

Zapata, R.; Azagra Malo, C. & M. Karlin. 2017. Tratamientos pregerminativos para la ruptura de la dormición en semillas de tres poblaciones de *Ramorinoa girolae*, leñosa endémica de zonas áridas en Argentina. *Bosque* 38(2): 237-246. [Pregerminative treatments for the rupture of dormancy in seeds of three populations of Ramorinoa girolae, woody endemic to arid zones in Argentina. *Forest*]

Karlin, U. O., Karlin, M. S., Zapata, R. M., Coirini, R. O., Contreras, A. M., & M. Carnero. 2017. La Provincia Fitogeográfica del Monte: límites territoriales y su representación. *Multequina*, (26), 63-75. [The Monte Fitogeográfica Province: territorial limits and their representation. *Multechin*]

## *Ulf Ola Torkel Karlin*

**Affiliation:** Universidad Nacional de Chilecito (UNdeC), La Rioja, Argentina.

**Education:** Agronomic Engineer. Facultad de Ciencias Agropecuarias. Universidad Nacional de Córdoba, Argentina. 1976.

**Research and Professional Experience:**
- Agroforestry Systems
- Nitrogen fixing trees
- Rural Development
- Arid Lands Management

**Director Research & Development Programs:**
- Founder Member and Program Director of the *Banco Nacional de Germoplasma de Prosopis* [National Germplasm Bank of *Prosopis*]. FCA-IFONA-FAO. 1985-89.
- Director of the Project *Manejo de Prosopis Arbóreos en Sistemas de Producción Agroforestales, Chaco Seco, Argentina* [Management of tree *Prosopis* in Agroforestry Systems, Arid Chaco, Argentina]. National Academy of Science (NAS). 1986-91.
- Coordinator of the *I-II-III Curso-Taller Internacional sobre Sistemas Agroforestales para Pequeños Productores en Zonas Áridas* [I-II-III International Course-Workshop about Agroforestry Systems for Smallholders in Arid Lands]. Facultad de Ciencias Agropecuarias, UNC. Villa Dolores, Córdoba. 1991-1992-1993.
- Expert Consultant in *El Avance de la Agroforestería en Zonas Áridas y Semiáridas de América Latina y el Caribe* [Advance of the Agroforestry in Arid and Semiarid Lands of Latin America and Caribean]. FAO. México 1993.
- Expert Consultant in *Manejo de Cuencas en Zonas Áridas y Semiáridas* [Basin Management in Arid and Semiarid Lands]. FAO. Argentina. 1994.
- International Consultor of the *International Foundation for Science* (IFS). Agroforestry Area. 1990-93 y 1995-98.

- Assessor and Trainer in the Program *Sistemas de Producción en Comunidades Aborígenes y de Pequeños Productores* [Production Systems in Communities of Aborigines and Smallholders]. Morillo, Chaco Salteño, GTZ. 1995-98.
- Consultant in *Development of Practical tools for participatory approaches to the National Action Programme. Processes to Combat Desertification*. Zambezi, N.E. Zimbabwe. PNUD. August 1996.
- Consultant in *Conocimientos y Tecnologías Tradicionales, sección América del Sur* [Knowledge and Tradicional Technologies, South America section]. PNUD. 1998.
- Technical Coordinator of the Project *Diagnóstico Bosques Nativos en el Chaco Argentino* [Native Woodland Diagnosis in the Argentine Chaco]. Sec. R.N. y D.S. de la Nación, Argentina – World Bank. 1998/99.
- Consultant for GTZ. *Programa Nacional de Lucha contra la Desertificación* [National Program to Combat Desertification]. La Paz, Bolivia. February-March 2000.
- Co-coordinator Project *Capacity Building in Dryland Development*. DAAD - National University of Nairobi, Kenia-Humboldt Universität, Berlín, Germany - UNC Córdoba, Argentina. 2005-2008.
- Professor and Co-coordinator of the Post degree Course *Sistemas Agroforestales en Zonas Áridas* [Agroforestry Systems in Arid Lands]. Escuela para Graduados. Facultad de Ciencias Agropecuarias – Universidad Nacional de Córdoba. March 2007.
- Co-coordinator of the Project *Manejo sustentable del ecosistema Salinas Grandes del Chaco* Árido [Sustainable Management of the Ecosystem Salinas Grandes of the Arid Chaco], Global Environment Facility, Argentina. 2007-2010.

- Director of the Project of *Manejo Sustentable y Participativo de los Bosques Nativos en el Valle del Bermejo, San Juan* [Sustainable and Participatory Management of Native Forests in the Bermejo Valley, San Juan]. Programa de Protección de los Bosques Nativos - Sec. Ambiente y Desarrollo Sustentable de la Nación. 2011-2016.
- Professor and Co-coordinator of the Post degree Course *Metodologías y Prácticas para la Multiplicación y el Manejo de Especies Aromáticas y Medicinales en Zonas Áridas* [Methodologies and Practices for the mutiplication and Management of Aromatic and Medicinal Plants in Arid Lands]. Universidad Nacional de San Juan. San Juan (UNSJ). April 2012.
- Post degree Professor of *Etnobotánica y Etnoecología* [Etnobotany and Ethnoecology]. UNSJ. August 2013.
- Post degree Professor of *Agroecología y Metodologías Participativas*. [Agroecology and Participatory Methodologies]. Universidad Nacional de La Pampa. November 2013.
- Post degree Professor of *Metodologías Participativas para el trabajo con Comunidades Rurales* [Participatory Methodologies for the work with Rural Communities]. UNSJ. August 2014 and August 2015.
- Post degree Professor of *Gestión y Manejo de Zonas Áridas*. [Management of Arid Lands]. UNdeC. November 2016.

**Professional Appointments:**
- Associate Professor and Coordinator of the Manejo de Agrosistemas Marginales [Management of Marginal Agrosystems] chair, Facultad de Ciencias Agropecuarias, Universidad Nacional Córdoba. 1985-2015.
- Founder y Member of Red Agroforestal Chaco [Agroforestry Network Chaco]. 1992-2018.
- Associate Professor and Coordinator of the Manejo de Bosques y Pasturas Naturales [Management of Forests and Natural Grasslands] chair, Departamento Biología, Universidad Nacional de San Juan. 2000-2015.

- Scientific Consultant Instituto de Ambientes de Montaña y Regiones Áridas (IAMRA) [Institute of Mountain and Arid Ranges]-Universidad Nacional de Chilecito, La Rioja, Argentina. 2015-2018

**Honors:** Doctor Honoris Causa in Agronomic Sciences. Faculty of Agronomic Sciences. National University of Córdoba, Argentina. 2014.

**Publications from the Last 3 Years:**

*Books:*

Demaio, P; Karlin, U. & M. Medina. 2015. *Árboles Nativos de Argentina. Tomo 1: Centro y Cuyo*. Ed. Ecoval. ISBN 978-987-45671-6-1. [*Native Trees of Argentina. Volume 1: Center and Whose.*]

Ojeda, M. & U. Karlin (Eds.). 2015. *Plantas Aromáticas y Medicinales. Modelos para su Domesticación, Producción y Usos Sustentables*. Ed. Universidad Nacional de Córdoba. ISBN 978-987-707-011-8. [*Aromatic and medicinal plants. Models for its Domestication, Production and Sustainable Uses*]

Demaio, P.; Karlin, U. & M. Medina. 2017. *Árboles Nativos de Argentina. Tomo 2: Patagonia*. Ed. Ecoval. ISBN 978-987-4003-16-4. [*Native Trees of Argentina. Volume 2: Patagonia.*]

Karlin, M. S. (ed.); Arnulphi, S. A; Karlin, U. O.; Bernasconi, J. R.; Accietto, R. y A. Cora. 2017. *Plantas del Centro de Argentina*. Ecoval Editorial. ISBN 978-987-4003-14-0. [*Plants of the Center of Argentina*]

*Book Chapters:*

Ojeda, M. & U. Karlin. 2015. Introducción a los yuyos. In: Ojeda, M. & U. Karlin (Eds.). 2015. *Plantas Aromáticas y Medicinales. Modelos para su Domesticación, Producción y Usos Sustentables*. Ed. Universidad Nacional de Córdoba. ISBN 978-987-707-011-8. Pp.: 13-20. [*Aromatic *and medicinal plants. Models for their Domestication, Production and Sustainable Uses.*]

Martinelli, M.; Karlin, U. O.; Inojosa, M.; Díaz, G. & I. Slavutzky. 2017. El bosque nativo en la región del Monte. In: Martinelli, M. & M. Inojosa (Eds.). *Los bosques del Monte: Conservación y Manejo de los Bienes Comunes Naturales*. Ed. Universidad Nacional de San Juan. ISBN 978-987-3984-42-6. Pp.: 43-70. [The native forest in the Monte region. In: *The forests of the Mount: Conservation and Management of Natural Common Goods.*]

Díaz, G. & U. O. Karlin. 2017. Especies leñosas. In: Martinelli, M. & M. Inojosa (Eds.). *Los bosques del Monte: Conservación y Manejo de los Bienes Comunes Naturales*. Ed. Universidad Nacional de San Juan. ISBN 978-987-3984-42-6. Pp.: 71-113. [Woody species. In: *The forests of the Mount: Conservation and Management of Natural Common Goods.*]

Karlin, U. O. & M. Karlin. 2017. Región Humedales valliserranos. 6a. Subregión Ríos y arroyos de los valles intermontanos. In: Benzaquén, L.; Blanco, D., Bo, R., Kandus, P., Lingua, G., Minotti, P. & R. Quintana (Eds.). *Regiones de Humedales de la Argentina*. Fundación para la Conservación y el Uso Sustentable de los Humedales-Wetlands International. Bs. As. ISBN 978-987-29811-6-7. Pp.: 163-172. [Vallliserran Wetlands Region. 6a. Subregion Rivers and streams of the intermontane valleys. In: *Wetlands Regions of Argentina*.]

Karlin, M. & U. O. Karlin. 2017. Región Humedales valliserranos. 6c. Subregión Salinas de la Depresión Central. In: Benzaquén, L.; Blanco, D., Bo, R., Kandus, P., Lingua, G., Minotti, P. & R. Quintana (Eds.). *Regiones de Humedales de la Argentina*. Fundación para la Conservación y el Uso Sustentable de los Humedales-Wetlands International. Bs. As. ISBN 978-987-29811-6-7. Pp.: 185-194. [Vallliserran Wetlands Region. 6c. Subregion Salinas of the Central Depression. In: *Wetlands Regions of Argentina*]

*Scientific Papers:*

Karlin, U. O., Karlin, M. S., Zapata, R. M., Coirini, R. O., Contreras, A. M., & M. Carnero. 2017. La Provincia Fitogeográfica del Monte: límites territoriales y su representación. *Multequina*, (26), 63-75. [*The Monte Fitogeográfica Province: territorial limits and their representation. Multechin*]

In: *Acacia*
Editor: Aide Matheson

ISBN: 978-1-53614-237-2
© 2018 Nova Science Publishers, Inc.

*Chapter 3*

# REGENERATION IN *ACACIA MANGIUM* WILLD. PLANTATION IN MT. MAKILING FOREST RESERVE, PHILIPPINES: IMPLICATIONS TO RESTORATION ECOLOGY

*Marilyn S. Combalicer*[1,*], *PhD,*
*Jonathan O. Hernandez*[1,†], *Pil Sun Park*[2,‡], *PhD*
*and Don Koo Lee*[2,§], *PhD*

[1]Department of Forest Biological Sciences,
College of Forestry and Natural Resources,
University of the Philippines Los Baños, Laguna, Philippines
[2]Department of Forest Sciences,
College of Agriculture and Life Sciences,
Seoul National University, Seoul, Republic of Korea

---

[*] Corresponding Author Email: mscombalicer@up.edu.ph.
[†] johernandez2@up.edu.ph.
[‡] pspark@snu.ac.kr.
[§] leedk@snu.ac.kr.

## Abstract

There have been a number of debates on the use of exotic species plantations especially for restoration in the Philippines, mainly due to some ecological concerns (e.g., loss of genetic variation). Hence, the potential of *Acacia mangium,* an exotic species, for restoration of a degraded land in Mt. Makiling Forest Reserve (MMFR), Philippines was examined. The compositional and structural changes of the grassland under *A. mangium* plantation over a 20-year and 28-year old study periods were analyzed. Results suggested a general trend of changes in *A. mangium* plantation which was once a grassland dominated by *Imperata cylindrica* and *Saccharum officinarum*. Both (stems ha$^{-1}$) and basal area (m$^2$ ha$^{-1}$) increased significantly in 2010-2018 ($P$=0.001). Increase in the dominance of native overstory trees coupled with a decrease in the dominance of exotic species was also observed. Lastly, results revealed an increase in species richness and species diversity with the presence of more native trees naturally occurring in MMFR (e.g., *Pterocymbium tinctorium and Celtis luzonica)*. These are all attributed to ecological processes including improved soil fertility and microclimate, disturbance, soil seed banks, seed flow of adjacent forests, seed dispersal, and mortality rate along with the N-fixing ability of *Acacia mangium.* This study, therefore, provides insights that *A. mangium* plantation is potential for restoration of degraded lands in the Philippines. However, this may require in-depth ecological knowledge, silvicultural management practices, and continuous field-based monitoring studies to enhance understanding of the management aspect of a restored landscape.

**Keywords:** exotic plantation, forest structure, grassland, native species, regeneration, species diversity

## Introduction

The natural forest loss across tropical countries, including the Philippines, has long been in serious decline (MacDicken et al., 2015, Parrotta et al., 1997). This is associated with greater loss of biodiversity, potential extinction of important species and reduced soil fertility. This will further lead to loss of capacity for regeneration and recovery due to soil degradation and disturbance on forest structure and composition. These impacts support the likelihood that restoration of degraded lands through natural regeneration is indeed a serious and difficult task. Recently, as the public becomes more concerned with the environment and biodiversity, the interest in the restoration of heavily degraded areas significantly increased (Hernandez et al., 2016; Lamb 2012; Combalicer et al., 2011; Montagini and Finney 2011; Newton and Tejedor 201). There have been many attempts of restoring degraded lands using fast growing species (Lamb and Tomlinson, 1994) and *A. mangium* is one of the commonly used species for plantations (de la Cruz, 1995). The potential of *A. mangium,* for instance, for landscape restoration was first recognized when it was introduced to *Imperata cylindrica* grasslands (Turnbull et al., 1998; Doran and Skelton, 1982). A number of studies revealed that when *Acacia* is planted into degraded ecosystem, recovery of favourable site conditions and recruitment of native seedlings were the important outcomes (Peng et al., 2005).

In the Philippines, however, there have been negative notions on the use of exotic species plantations for any purpose of greening, more so on restoration. This is particularly because these plantations are often monocultures which raise concerns about ecological conservation (e.g., loss of genetic variation) (Kelty, 2006). In the country, we had a century of experience of using exotic species, but these are generally for reforestation and/or rehabilitation not for restoration. Hence, attempts to restore degraded lands in the country using exotic species (e.g., *A. mangium*) will require science-based information to ensure its success. Further, one limitation pertaining to the use of exotic species for restoration is the lack of access to field-based information specifically about implications to restoration ecology. As the objective of this study, the potential of *A. mangium* plantation for restoration of a degraded land was investigated by analyzing the compositional and structural changes over a 20-year and 28 year old study periods.

## MATERIALS AND METHODS

### Study Area

The Mt. Makiling Forest Reserve (MMFR) is an ASEAN Heritage Park. It is located at 14° 8' north and 121° 12' east, approximately 65 km southeast of Metro Manila. The forest reserve has a total area of 4,244 hectares and is one of the centers of plant diversity in the Philippines. The climate is tropical monsoon with short dry season. Mean annual precipitation and temperature are 24°C and 2, 397 mm, respectively (Han et al., 2010). The dominant soil is clay loam derive from volcanic tuff with andesite and a basalt base.

There are four vegetation types in the forest reserve; namely, mossy forest zone, dipterocarp mid-montane forest zone, grassland, and agroforestry zone (Gruezo, 1997). Of these vegetation types, the sampling plots were situated at the grassland zone in Sitio Kay Inglesia at altitude ranging from 500 to 800masl (Abraham et al., 2007). This area had been previously cultivated and perennially burned prior to the 1990s and was previously covered by *Saccharum officinarum* L. and *Imperata cylindrica* L (Figure 1).

### Species Structure and Composition

In 2009 and 2018, three 20m x 20m sampling plots were established in the study site (i.e., 28-year old plantation). Trees with a diameter at breast height (DBH) above 3 cm were identified. Species nomenclature follows the correct and accepted name in The Plant List Database (a working list of all plant species) and Co's Digital Flora of the Philippines. Their DBH and height were measured using diameter tape and meter pole, respectively. The DBH was measured at a height of 1.3m above the ground. Inside the plots, identification of saplings and seedlings were also conducted.

*Regeneration in* Acacia Mangium *Willd ...*

a

b

Source: J. M. Maloles and M. S. Combalicer.

Figure 1. The study site in Sitio Kay Inglesia, MMFR showing (a) grassland dominated by *S. officinarum* and *I. cylindrica* in 1993 and (b) the same grassland planted with *A. mangium* in 2003.

## Data Preparation and Analysis

For each study period interval (i.e., 1990-2009 and 2010-2018), all trees were grouped into two categories: (i) saplings (≥ 3 cm - 9.9 cm dbh) and (ii) overstory stems (≥10 cm). This was done for examination of species structure and mortality. For all categories, the density (stems ha$^{-1}$) and basal area (m$^2$ ha$^{-1}$) by species were computed. Changes in means of stem density between study periods were compared with paired $t$-tests at α=0.05 while Wilcoxon signed rank tests for basal area (BA) using R-Studio Statistical Package software version 3.2.2. The calculation of relative density (total density of species on a plot/total density of all species on a plot) and relative basal area (total BA of species on a plot/total BA of all species on a plot) followed. The values derived were used to compute for importance values following the formula in Lowney et al., (2015). We calculated the annual mortality rate (AMR) following the modified formula in Lutz and Halpern (2006) for both saplings and overstory stems. The species richness (S), Shannon-Weiner diversity (H') and evenness (E) were calculated using Mutivariate Statistical Package (MVSP) software version 3.22. Lastly, the list of new recruited native species across plots along with their geographic range were also prepared.

# RESULTS

## Stand Density

In 2010-2018, over 150 live stems from 25 different species were recorded from the three plots, compared to 120 live stems from only 12 species in 1990-2009. The count of stems per hectare in both sample intervals varied significantly (Figure 2). The stem density (i.e., 3 cm to 10 cm dbh) in 1990-2009 ranged from 25 to 225 stems ha$^{-1}$ with a mean density of 67.39±11.60. Live overstory stems (i.e., ≥ 10 cm dbh) are more compacted, ranging from 75 to 880 stems ha$^{-1}$ with a mean density of

371±41.28. As shown in Table 1, a significant increase in the number of stems ha$^{-1}$ (i.e., 3 cm to 10 cm dbh) was observed in 2010-2018 year old stand ($P$=0.008) which is about 7%. This also holds true to stems with dbh greater than 10 cm (i.e., $P$=0.001 at about 3%). In 2010-2018 sample interval, the species with the highest stem density, other than *A. mangium*, is *Syzygium simile* (250 stems ha$^{-1}$) compared to *S. macrophylla* (400 stems ha$^{-1}$) in 1990-2009. This is followed by *Neonauclea bartlingii* (DC.) Merr. (175 stems ha$^{-1}$) and *Ficus nota* (Blanco) Merr. (150 stems ha$^{-1}$).

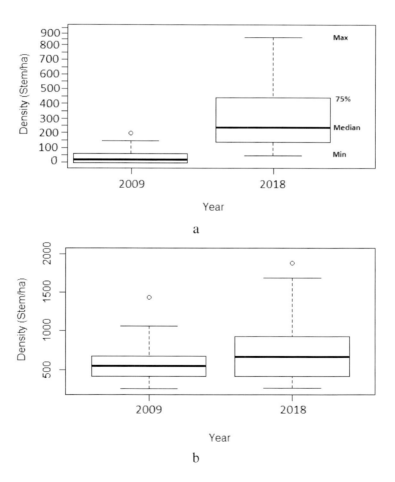

Figure 2. Change in density of (stems ha$^{-1}$) of (a) saplings and (b) overstory stems over a 20-year and 28-year old study periods.

**Table 1. Changes in means (± SE) density (stems ha⁻¹) of all stems in two diameter classes (3 cm - 9.9 dbh and ≥10 cm dbh)**

| Diameter Classes | 2009 | 2018 |
| --- | --- | --- |
| Saplings (3 cm - 9.9 cm dbh) | 67.39±11.60 | 371.04±41.28 |
| % Δ 1990-2009 and 2010-2018 (*P* value) | -7.0817(0.008) | |
| Overstory (≥10 cm dbh) | 578.22±30.88 | 721.25±44.29 |
| % Δ 1990-2009 and 2010-2018 (*P* value) | -2.66 (0.001) | |

Means between sample intervals were compared with paired *t*-tests at α=0.05.

## Basal Area and Importance Values

Along with stem density, changes in BA of all tree species emphasized compositional change of the forest (Figure 3). In both sample intervals, the BA of common tree species in Mt. Makiling increased significantly with p-value ranging from 0.001 to 0.016, except Leucaena leucocephala (Lam.) de Wit (p=0.089). The difference in BA of 23.43 m² ha⁻¹ (t = -10.236, p < 0.001) was observed in S. simile which also holds true to its increase in density in 2010-2018. The changes in both BA and density of the common species brought also a significant change in their importance values (IV) as presented in Figure 4. For example, the increase in both BA and density of S. simile drove an increase in importance value (i.e., 8.03% from only 5.76) in the present study period. This is followed by *N. bartlingii* with difference of 18.87 m² ha⁻¹(t = -5.517, p=0.001), resulting to an increase in its importance value (IV) of 14.63% from only 6.24% in 1990-2009. On the other hand, the low increase in BA was more pronounced in *A. mangium* with difference of only 10.280 m² ha⁻¹ (t = -3.711, p < 0.003). This slow increase in BA corresponded to decrease in *A. mangium* stem density in 2010-2018 sample interval. This resulted to a general decline in its IV (i.e., 25% from 31%). In spite of this, *A. mangium* is still the most dominant species in the area followed by Ficus septica Blanco (23.09%) and *Pterocarpus indicus* Willd (15.83%). A decline in the BA, density, and IV of *Calliandra calothyrsus* Meissner, *L. leucocephala*, and *Swietenia macrophylla* King was observed, making them the least dominant in the area.

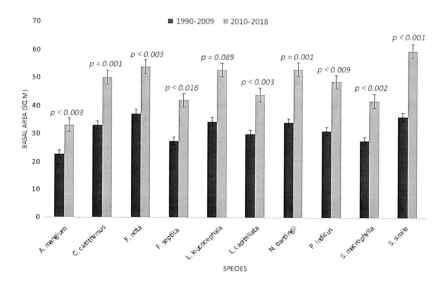

Figure 3. Changes in mean (± SE) basal area (m² ha⁻¹) of ten common tree species (≥3 -0-9.9 cm dbh) in Mt. Makiling Forest Reserve between 1990-2009 and 2010-2018. Means between sample intervals were compared with Wilcoxon signed rank tests at α=0.05.

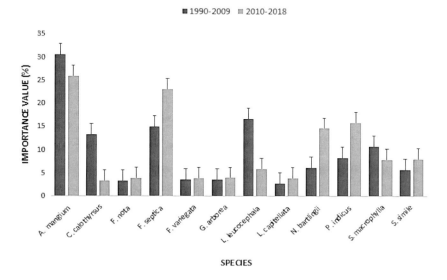

Figure 4. Changes in importance values of common tree species in the sampled forest in Mt. Makiling Forest Reserve between 1990-2009 and 2010-2018.

## Mortality

Annual mortality rate (AMR) differed between size classes in both sample intervals (Table 2). There are more stems died in the forest in 1990-2009 than 2010-2018. In the 20-year old study period, the total AMR of overstory stems was 96% and 80% for the saplings. Between 2010 and 2018, the AMR of only 72% and 74% were observed for the overstory stems and saplings respectively.

Figure 5. Leaning and decaying boles of *A. mangium* in Sitio Kay Inglesia Mt. Makiling Forest Reserve.

**Table 2. Changes in % annual mortality rate (AMR) of all stems in two diameter classes (3 cm - 9.9 cm dbh and ≥10 cm dbh) between 1990-2009 and 2010-2018 year old forest**

| Diameter class | AMR 1990-2009 (%) | AMR 2010-2018 (%) |
|---|---|---|
| Saplings (3 cm - 9.9 cm dbh a | 80 | 74 |
| Overstory stems (≥10 cm dbh) | 96 | 72 |

Among stems, ≥10 cm dbh, mortality was highest for *A. mangium* (Figure 5), which experienced 85% AMR between the two study periods. The lowest AMR=78% was observed in *P. indicus*. The *S. simile* and *N. bartlingii* experienced 0% AMR.

## Species Richness, Evenness and Diversity

In all stems, results revealed significant increase in both species richness and diversity in 2010-2018 study period (Table 3). A total of 25 species are present in the forest from only 12 species in 1990-2009. Twelve out of 13 new species are indigenous in the Philippines, of which 23% is endemic to Luzon, Philippines: namely, *Diplodiscus paniculatus* Turcz, *Gongrospermum philippinense* Radlk, and *Strombosia philippinensis* (Baill.) vid (Table 4). All these new species are at sampling stage with mean dbh ranging from 4.27±2.21 to 10.15±1.77. In terms of evenness, there is no significant change observed between the sample intervals.

Evidence of active regeneration on the ground layer was also observed inside the three plots through the presence of seedlings (*c.a* 2 cm-5 cm dbh) of various tree species. Most of these seedlings are also native to the Philippines. These species include: *Ardisia pyramidalis* (Cav.) Pers., *Mimusops elengi* Linn., *Neonauclea media* (Havil.) Merr., *Canthium monstrosum* (A.Rich.) Merr., *Tabernaemontana pandacaqui* Poir., *Sterculia oblongata* R.Br., *Pterocymbium tinctorium* Merr., *Dysoxylum gaudichaudianum* (A. Juss.) Miq., *Melanolepis multiglandulosa* (Reinw. ex Blume) Reichb. & Zoll., *Turpinia ovalifolia* Elmer, *Cratoxylon sumatranum* (Jack) Blume, *Aglaia tomentosa* Teysm. & Binnend., *Alstonia macrophylla* Wallich ex G. Don, *Scolopia luzonensis* (Presl.) Warb., *Alangium chinense*

(Loureiro) Harms, *Flacourtia indica* (Burm. f.) Merr., *Mallotus cumingii* Müll.Arg., *Ardisia crenata* Sims., *Clerodendrum intermedium* Cham., and *Neonauclea formicaria* (Elmer). It was also seen the presence of *Celtis luzonica* Warb across plots.

**Table 3. Changes in species richness, evenness, Shannon-Weiner diversity of all stems between 1990-2009 and 2010-2018**

| Index | 1990-2009 | 2010-2018 |
|---|---|---|
| Richness (S) | 12 | 25 |
| Evenness (E) | 0.759 | 0.722 |
| Shannon-Weiner diversity (H') | 1.885 | 2.832 |

**Table 4. List of newly recruited indigenous tree species in the forest in 2009-2018 study period**

| Species | DBH Mean ± SE | Geographic range |
|---|---|---|
| *Antidesma pentandrum* (Blanco) Merr. | 7.10±0.22 | In |
| *Artocarpus odoratissimus* Blanco | 7.95±0.77 | Ex |
| *Breynia rhamnoides* (Retz.) Muell.-Arg. | 6.50±0.22 | In |
| *Cratoxylum formosum* Dyer | 8.53±2.02 | In |
| *Crypteronia paniculata* Blume | 10.15±1.77 | In |
| *Diplodiscus paniculatus* Turcz | 7.40±1.11 | En |
| *Gongrospermum philippinense* Radlk | 4.27±2.21 | En |
| *Leucosyke capitellata* Wedd. | 6.2±1.27 | In |
| *Litsea elliptica* Blume | 5.4±1.21 | In |
| *Neonauclea bartlingii* (DC.) Merr. | 6.86±0.89 | In |
| *Polyscias nodosa* (Blume) Seemann | 7.25±1.06 | In |
| *Strombosia philippinensis* (Baill.) vid | 7.1±0.84 | En |
| *Wendlandia uvariifolia* Hance | 10.2±1.70 | In |

In, Ex, En denote indigenous, exotic and endemic, respectively.

# DISCUSSION

The results suggest a general trend of compositional and structural changes in *Acacia mangium* plantation which was once a grassland dominated by *Imperata cylindrica* (Cogon) and *Saccharum officinarum*. We found out that *A. mangium* may be helpful for forest restoration. Its potential for restoring degraded lands because of its ability to tolerate wide range of environments have long been established (Doran and Skelton, 1982). In this study, the increase in sapling density (i.e., ≤ 2m in height) of indigenous species such as *Syzgium simile* and *N. bartlingi* implies a productive natural recruitment process in the area. The dispersal and recruitment processes are major determinant of species abundance (Daling et al., 1998). Changes in forest species composition are also influenced by a number of factors including recruitment from soil seed banks (Burton et al., 2011; Lovett et al., 2006). This is possible in the absence of adult native vegetation serving as both seed source and nurse habitats (Yelenik, 2014). *Syzgium simile* and *N. bartlingii* are native in the Philippines, particularly at low to medium altitudes (Merrill, 1923) and are naturally occurring in Mt. Makiling. Consequently, the increased density of their saplings in the area may be attributed to germination of their seeds which might have been deposited in soil long time ago prior to degradation. The microclimate within the plantation may have significantly improved with the planting of *A. mangium,* favoring the germination of native seeds that were originally present in the area. Improvement of microclimate and soil enzyme activity under *Acacia* plantations has been reported (Lee et al., 2006). This also holds true to other tree plantation species used in the restoration of degraded lands (Carnus et al., 2006; Parrotta et al., 1997). The present study conforms to the result revealed in Combalicer et al., (2011) in which increase in density of woody species (> 5 cm dbh) was attributed to intensity of competition for resources (e.g., nutrients, water, space, and light) available for plant use. In addition, however, adjacent forests which may act as possible source of seeds along with dispersal processes can also be a contributing factor of the increased stem density of most indigenous species across plots.

The basal area of tree species common to both sample intervals also indicated compositional change of the plantation. The BA of both exotic and native species significantly increased which implies a stability of BA in the forest despite the observed decline in stem density of overstory trees (≥ 3m in height). Tulod et al., (2017) reported the same higher BA of exotic species in association with native ones in Southern, Philippines. The disturbance brought by the typhoon Glenda (*Rammasun*) in 2014 may also explain the increase of BA as well as stem density. As a result of disturbance, the plantation might have experienced a rapid regeneration and canopy recruitment due to increased growing space. Because of improved light condition, reduced root competition and increased nutrient availability in soil, and decreased air humidity, species compositional change may depend on gap phase generation (Brown 1994). Oliver and Larson (1996) explained how the existing and new established stems in an area may take advantage of the gap created by a disturbance. The same result was observed in the study of Sagar et al., (2003) and Ramirez-Marcial et al., (2001). A positive correlation between BA and rate disturbance was also reported in Smiet (1992).

As the stem density and BA changed, the IV of tree species within the plantation also changed. This suggests that all species (e.g., *S. simile* and *N. bartlingii*) that increased in both density and BA are expected to have increasing IV, thereby changing their dominance in the area. The new set of dominant species with their respective IV may suggest different habitat preference, mode of adaptation and site requirements. Although it showed a low increase in the BA, the decline in stem density of *A. mangium* drove a decrease in its IV. The BA may not be sufficient to compensate for the loss of its standing stems which may be due to its high AMR. Abrupt decline in density of the species may be observed after the disturbance by exposing the overstory stems (≥ 3m in height) to destructive winds of the typhoon Glenda (*Rammasun*). This resulted to increased debarking and rotting of boles of *A. mangium* which are evident in the plantation within 2010-2018 study period. In spite of these, *A. mangium* is still the dominant species which may be the function of its natural defense and mechanisms against disturbance. The species are well-adapted to harsh environments (Brockwell et al., 2005); hence, it is usually selected as pioneer species for restoring degraded lands (Yang et al., 2009; Nichols and Carpenter, 2006). According to Combalicer et al. (2012), *A. mangium* could survive in various environments in view of its physiological attributes (i.e., high net photosynthesis rate and photosynthetic nitrogen utilization efficiency). Further, *Acacia mangium* has $N_2$-fixation ability that can increase N content in soil (Forrester et al., 2006).

This has long been associated with positive growth rate (Bouillet et al., 2008; Forrester et al., 2006; Brockwell et al., 2005).

There were more stems died in 1990-2009 than 2010-2018 study periods. This indicated that the transition of forest structure development was already underway in the plantation. This is considering that the plantation was once characterized by poor soil productivity typical of a cogonal grassland. The area had also been perennially burned prior to the 1990s (Lee *et al.,* 2006). Consequently, varying preference and requirement across species may continue to escalate as forest development progresses through natural regeneration (Pardos et al., 2005). However, the colonization of woody species undergoes unpredictable and dynamic ecological processes (Ruiz-Jaen and Aide, 2005), such that only the resilient and adaptive species shall thrive and colonize the area. Every species in the plantation has varying preferences, thereby eliminating the unfit ones. This can be shown by the lowest AMR observed in *P. indicus, S. simile* and *N. bartlingii* which are all native trees in the area. *Pterocarpus indicus* is a leguminous tree that can fix atmospheric nitrogen (Binkley and Giardina, 1997) essential for their growth under an area undergoing restoration. This association present in leguminous species in the plantation, including *A. mangium* (Cole et al., 1996), may have facilitated the emergence of additional native species by enhancing soil fertility. Enhanced soil fertility may eventually lead to increased resource availability. According to Franco and De Faria (1997), legume tree species can enhance soil organic carbon (SOC) up to 12 mg dry litter ha$^{-1}$ year$^{-1}$. In *A. mangium*, the recorded rate of carbon deposition ranges from 9.4 to 11.1 mg ha$^{-1}$ year$^{-1}$ (Hardiyanto and Wicaksono, 2008). Further, the sample plantation in Sitio Kay Inglesia has clayey soil type (Lee *et al.*, 2006). Records show that SOC in clay soil is associated with productive *Acacia* plantation (Korschens et al. 1998).

The increase in species diversity (i.e., low to moderate) and species richness in the plantation emphasized the changes that occurred between 1990-2009 and 2010-2018. The same result was observed in Tulod et al., (2017). Interestingly, the occurrence of a critically endangered and endemic *D. paniculatus* also implies that exotic plantation can be as favorable habitat as the native forest in association with the other native species. It was also observed the presence of seedlings of *C. luzonica*, an endemic species to the Philippines (Word Conservation Monitoring Center, 1998). *P. tinctorium* and *C. luzonica* are dominant species in Mt. Makiling (Han et al., 2012; Galang, 2010). This emergence of new native and endemic species, which were not reported in 1990-2009 study period (Lee e*t al.*, 2006), is a good measure of a successful restoration (Ruiz-Jaen and Aide, 2005). This further suggests that a gradual shift in the landscape structure from grassland to *Acacia* plantation toward a restored forest is already in progress. This is because the species recorded were mostly either pioneer or intermediate species which are common species in an area undergoing succession. Examples of these species are *F. septica, F. nota, M. multiglandulosa, A. macrophylla, Dysoxylum gaudichaudianum,* and *Tabernaemontana pandacaqui*.

## CONCLUSION

The regeneration and/or recruitment of native tree species in a degraded land dominated by *I. cylindrica* is possible under *Acacia magium* plantation. It can be a nurse crop for bringing back native tree species originally present in the site. The composition, structure, soil condition, and microclimate of the site also showed significant improvement under such plantation. This study, therefore, provides insights that *A. mangium* plantation is potential for restoration of degraded lands in the Philippines. The present insights, however, may require in-depth ecological knowledge, silvicultural management practices, and continuous field-based monitoring studies to enhance understanding of the management aspect of a restored landscape.

## REFERENCES

Binkley, D. & Giardina, C. (1997). Nitrogen fixation in tropical forest plantations. In: Nambiar, E. K. S., Brown, A. G. (Eds.), *Management of soil, nutrients and water in tropical plantation forests.* ACIAR, CSIRO, CIFOR, Canberra, pp. 297–337.

Bouillet, J., Laclau, J., Gonçalves, J., Moreira, M., Trivelin, P., Jourdan, C. & Galiana, A. (2008). Mixed-species plantations of *Acacia* mangium and Eucalyptus grandis in Brazil. *Forest Ecology and Management, 255*(12), 3918-3930. doi:10.1016/j.foreco.2007.10.050.

Brockwell, J., Searle, S. D., Jeavons, A. C., & Waayers, M. (2005). *Nitrogen fixation in Acacias: An untapped resource for sustainable plantations, farm forestry and land reclamation.* Canberra: Australian Centre for International Agricultural Research.

Brown, N. (1993). The implications of climate and gap microclimate for seedling growth conditions in a Bornean lowland rain forest. *Journal of Tropical Ecology, 9*(02), 153-168. doi: 10.1017/s0266467400007136.

Burton, J. I., Mladenoff, D. J., Clayton, M. K., & Forrester, J. A. (2011). The roles of environmental filtering and colonization in the fine-scale spatial patterning of ground-layer plant communities in north temperate deciduous forests. *Journal of Ecology, 99*(3), 764-776. doi: 10.1111/j.1365-2745.2011.01807.x.

Carnus, J. M., Parrotta, J., Brockerhoff, E. G., Arbez, M., Jactel, H., Kremer, A., Lamb, D., O'Hara, K. & Walters, B. (2006). Planted forests and biodiversity. *Journal of Forestry, 104*(1), 65–77.

Cole, T. G., Yost, R. S., Kablan, R., & Olsen, T. (1996). Growth potential of twelve *Acacia* species on acid soils in Hawaii. *Forest Ecology and Management, 80*(1-3), 175-186. doi: 10.1016/0378-1127(95)03610-5.

Combalicer, M. S., Lee, D. K., Woo, S. Y., Hyun, J. O., Park, Y. D., Lee, Y. K., Combalicer, E. A., & Tolentino, E. L. (2012). Physiological Characteristics of *Acacia* auriculiformis A. Cunn. ex Benth., *Acacia* mangium Willd. and *Pterocarpus indicus* Willd. in the La Mesa Watershed and Mt. Makiling, Philippines. *Journal of Environmental Science and Management, (1)*, 14-28.

Combalicer, M. S. (2011). Aboveground biomass and productivity of nitrogen-fixing tree species in the Philippines. *Scientific Research and Essays, 6*(27). doi:10.5897/sre11.1633.

Dalling, J. W., Hubbell, S. P., & Silvera, K. (1998). Seed dispersal, seedling establishment and gap partitioning among tropical pioneer trees. *Journal of Ecology, 86*(4), 674-689. doi: 10.1046/j.1365-2745.1998.00298.x.

De la Cruz R. E. (1995). *Past, Present and Future Trends in Reforestation Research in the Philippines*. P6.11-00 Forest Sector Analysis. Bio-Reafforestation in the Asia-Pacific Region.

Doran, J. C. & Skelton, D. J. (1982). *Acacia mangium* seed collections for international provenance trials. In. *FAO*. http://agris.fao.org/agris-search/search.do?recordID=XF19830900878

Forrester, D. I., Bauhus, J., Cowie, A. L., & Vanclay, J. K. (2006). Mixed-species plantations of *Eucalyptus* with nitrogen-fixing trees: A review. *Forest Ecology and Management, 233*(2-3), 211-230. doi: 10.1016/j.foreco.2006.05.012.

Franco, A. A., & Faria, S. M. (1997). The contribution of N2-fixing tree legumes to land reclamation and sustainability in the tropics. *Soil Biology and Biochemistry, 29*(5-6), 897-903. doi: 10.1016/s0038-0717(96)00229-5.

Gruezo, W. S. (1997). Floral diversity profile of Mount Makiling Forest Reserve, Luzon, Philippines. In: Dove, M.R. and P.E. Sajise (eds.). *The Condition of Biodiversity Maintenance in Asia: The Policy Linkage between Environmental and Sustainable Development*. Khon Kaen and East-West Center, Honolulu. pp. 153-175.

Han, S. R., Woo, S. Y., & Lee, D. K. (2010). Carbon storage and fl ux in aboveground vegetation and soil of sixty-year old secondary natural forest and large leafed mahogany (*Swietenia macrophylla* King) plantation in Mt. Makiling, Philippines. *Asia Life Sciences*, *19*(2), 357-372.

Han, A. R., Sohng, J. E., Barile, J. R., Lee, Y. K. Woo, S. Y., Lee, D. K., & Park, P. S. (2012).Comparison of Soil Seed Banks in Canopy Gap and Closed Canopy Areas between a Secondary Natural Forest and a Big Leaf Mahogany (*Swietenia macrophylla* King) Plantation in the Mt. Makiling Forest Reserve, Philippines. *Journal of Environmental Science and Management*, 4(1), 47-59.

Hardiyanto, E. B. & Wicaksono, A. (2008). Inter-rotation site management, stand growth and soil properties in *Acacia* mangium plantations in South Sumatra, Indonesia. In: Nambiar, E. K. S. (Ed.), *Site management and productivity in tropical plantation forests.* CIFOR, Piracicaba, Brazil and Bogor, Indonesia, pp. 107–122.

Hernandez, J. O., Malabrigo Jr., P. L., Quimado, M. O., Maldia, L. S. J., & Fernando, E. S., (2016). Xerophytic characteristics of *Tectona philippinensis* Benth. & Hook. *Philippine Journal of Science, 145*(3): 259–269.

Kelty, M. J. (2006). The role of species mixtures in plantation forestry. *Forest Ecology and Management, 233*(2-3), 195-204. doi:10.1016/j.foreco.2006.05.011.

Körschens, M., Weigel, A., & Schulz, E. (1998). Turnover of soil organic matter (SOM) and long-term balances - tools for evaluating sustainable productivity of soils. *Zeitschrift Für Pflanzenernährung Und Bodenkunde, 161*(4), 409-424. doi:10.1002/jpln.1998.358161040.

Lamb, D. (2012). Forest restoration – the third big silvicultural challenge. *Journal of Tropical Forest Science, 24* (1): 121-129.

Lee, Y. K., Lee, D. K., Woo, S. Y., Park, P. S., Jang, Y. H., & Abraham, E. R. (2006). Effect of *Acacia* plantations on net photosynthesis, tree species composition, soil enzyme activities, and microclimate on Mt. Makiling. *Photosynthetica, 44*(2), 299-308. doi: 10.1007/s11099-006-0022-9.

Lovett, G. M., Canham, C. D., Arthur, M. A., Weathers, K. C., & Fitzhugh, R. D. (2006). Forest Ecosystem Responses to Exotic Pests and Pathogens in Eastern North America. *BioScience, 56*(5), 395. doi: 10.1641/0006-3568(2006)056[0395:fertep]2.0.co;2.

Lowney, C. A., Graham, B. D., Spetich, M. A., Shifley, S. R., Saunders, M. R., & Jenkins, M. A. (2015). Two decades of compositional and structural change in deciduous old-growth forests of Indiana, USA. *Journal of Plant Ecology, 9*(3), 256-271. doi: 10.1093/jpe/rtv050.

Lutz, J. A., & Halpern, C. B. (2006). Tree Mortality During Early Forest Development: A Long-Term Study Of Rates, Causes, And Consequences. *Ecological Monographs, 76*(2), 257-275. doi: 10.1890/0012-9615(2006)076[0257:tmdefd]2.0.co;2.

Macdicken, K. G. (2015). Global Forest Resources Assessment 2015: What, why and how? *Forest Ecology and Management, 352*, 3-8. doi: 10.1016/j.foreco.2015.02.006.

Marsoem, S. N., & Irawati, D. (2016*). Basic properties of Acacia mangium and Acacia auriculiformis as a heating fuel*. doi:10.1063/1.4958551.

Merrill, E. D. (1922). *An enumeration of Philippine flowering plants*. doi:10.5962/bhl.title.49412.

Montagnini, F., & Finney, C. (2011). *Restoring Degraded Landscapes with Native Species in Latin America*. New York: Nova Science.

Newton, A. C., & Tejedor, N. (2011). *Principles and practice of forest landscape restoration: Case studies from the drylands of Latin America*. Gland, Switzerland: IUCN.

Nichols, J. D., & Carpenter, F. L. (2006). Interplanting Inga edulis yields nitrogen benefits to *Terminalia amazonia*. *Forest Ecology and Management, 233*(2-3), 344-351. doi:10.1016/j.foreco.2006.05.031.

Oliver, C. D., & Larson, B. C. (1996). *Forest stand dynamics*. New York: John Wiley & Sons.

Pardos, M., Castillo, J. R., Cañellas, I., & Montero, G. (2005). Ecophysiology of natural regeneration of forest stands in Spain. *Investigación Agraria: Sistemas Y Recursos Forestales, 14*(3), 434. doi: 10.5424/srf/2005143-00939. [Ecophysiology of natural regeneration of forest stands in Spain. *Agricultural Research: Forest Systems and Resources*]

Parrotta, J. A. (1997). *Catalyzing native forest regeneration on degraded tropical lands: Selected papers based on the proceeding of an international symposium and workshop held in Washington, D.C., June 11-14, 1996*. Amsterdam: Elsevier Science.

Peng, S.L., Liu, J. & Lu, H. F. (2005). Characteristics and role of *Acacia* auriculiformis on vegetation restoration in lower subtropics of China. *Journal of Tropical Forest Science, 17*(1), 508-525.

Ramirez-Marcial, N., González-Espinosa, M., & Williams-Linera, G. (2001). Anthropogenic disturbance and tree diversity in Montane Rain Forests in Chiapas, Mexico. *Forest Ecology and Management, 154*(1-2), 311-326. doi: 10.1016/s0378-1127(00)00639-3.

Ruiz-Jaen, M. C., & Aide, T. M. (2005). Restoration Success: How Is It Being Measured? *Restoration Ecology, 13*(3), 569-577. doi: 10.1111/j.1526-100x.2005.00072.x.

Sagar, R., Raghubanshi, A., & Singh, J. (2003). Tree species composition, dispersion and diversity along a disturbance gradient in a dry tropical forest region of India. *Forest Ecology and Management, 186*(1-3), 61-71. doi: 10.1016/s0378-1127(03)00235-4.

Smiet, A. C. (1992). Forest ecology on Java: Human impact and vegetation of montane forest. *Journal of Tropical Ecology, 8*(02), 129-152. doi: 10.1017/s026646740000626x

Tulod, A. M., Casas, J. V., Marin, R. A., & Ejoc, J. A. (2017). Diversity of native woody regeneration in exotic tree plantations and natural forest in Southern Philippines. *Forest Science and Technology, 13*(1), 31-40. doi:10.1080/21580103.2017.1292958.

Turnbull, J. W., Midgley, S. J. & Cossalter, C. (1998). Tropical *Acacias* planted in Asia: an overview. In, *Recent developments in Acacia planting*. ACIAR, Hanoi, pp. 14–28.

World Conservation Monitoring Centre. (1998). *Celtis luzonica*. *The IUCN Red List of Threatened Species 1998*. http://dx.doi.org/10A9769709.en. Downloaded on 29 April 2018.

Yang, L., Liu, N., Ren, H., & Wang, J. (2009). Facilitation by two exotic *Acacia*: *Acacia* auriculiformis and *Acacia* mangium as nurse plants in South China. *Forest Ecology and Management, 257*(8), 1786-1793. doi: 10.1016/j.foreco.2009.01.033.

Yelenik, S. G., Dimanno, N., & Dantonio, C. M. (2014). Evaluating nurse plants for restoring native woody species to degraded subtropical woodlands. *Ecology and Evolution, 5*(2), 300-313. doi: 10.1002/ece3.1294.

## BIOGRAPHICAL SKETCHES

### *Marilyn S. Combalicer*

**Affiliation:** Department of Forest Biological Sciences, College of Forestry and Natural Resources, University of the Philippines Los Banos

**Education:** Dr. Combalicer earned her Master's and PhD degrees in forest sciences major in ecophysiology in 2004 and 2011, both from Seoul National University (SNU).

**Research and Professional Experience:**
- Dr. Combalicer is currently an Assistant Professor in the field of ecophysiology.
- Dr. Combalicer is one of the study leaders in the ASEAN-Korea Environmental Cooperation Project (AKECOP) in the Philippines since 2013. She is also involved in the researches titled *"Elucidating the Population Genetic Structure of Philippine Teak (Tectona philippinensis* Benth. & Hook, *Lamiaceae) for its Genetic Conservation and Management"* from 2018-2019 and *"Germplasm Conservation of Select Indigenous Forest Tree Species in Mt. Makiling Forest Reserve"* from 2016-2019.
- She worked as Programme Consultant in the International Cooperation Division, International Affairs Bureau, Korea Forest Service in 2012.

**Professional Appointments:** Dr. Combalicer served as member of the evaluating team of the ASEAN Forest Cooperation (AFoCo) projects in Brunei Darussalam, Indonesia, Malaysia and Philippines as well as member of the evaluating team of the AKECOP projects in ASEAN region.

She became the Vice President of the Forests and Natural Resources Society of the Philippines, Inc. in 2014 and is a member of the UPLB Gamma Sigma Delta Honor Society of Agriculture.

**Honors:**
- Dr. Combalicer was a recipient of the College of Forestry and Natural Resources (CFNR) Outstanding Teacher Award under Senior Category in the field of Forest Biological Sciences during the 106th CFNR Alumni Homecoming on14 April 2016. She was also part of the team who won the UPLB Outstanding Research Team in 2015.
- She became the recipient of the Graduate Scholarship for Excellent Foreign Students in SNU from 2007 up to 2009, Seoul National University Scholarship for Forest Environment Leadership Program of the Alumni in 2008, and AKECOP Scholarship in 2001-2003.

**Recent Publication:** She published several articles in ISI journals. The latest is chapter in a book:

Lee, D.K., M.N. Salleh, W.M. Ho, and M.S. Combalicer. 2016. Changing Trends of Forestry Research Demand. Eds. Pancel, L. and M. Kohl. *Tropical Forestry Handbook*. Second Edition. Springer-Verlag Berlin Heidelberg. pp. 3559 -3570.

### *Jonathan Ogayon Hernandez*

**Affiliation:** Department of Forest Biological Sciences, College of Forestry and Natural Resources, University of the Philippines Los Baños (UPLB)

**Education:** For. Hernandez earned his Bachelor's degree in Forestry in 2016 from UPLB, where he chose environmental forestry as his specialization.

**Research and Professional Experience:** For. Hernandez has been with UPLB since 2016 working as an instructor. He handles laboratory classes such as forest botany, forest ecology, plant physiology, forest biodiversity, and forest plants taxonomy. His research interests are forest restoration, biodiversity, ecophysiology and forest plants anatomy. Recently, For.

Hernandez is a member of a project entitled "Elucidating the population genetic structure of Philippine Teak (*Tectona philippinensis* Beth. & Hook, Lamiaceae) for its Genetic Conservation and Management."

In 2017, he was able to present (oral) his paper in the 12th International Congress of Ecology (INTECOL) of International Association of Ecology held in Beijing China. The title of his paper was: "Coastal Forest Restoration Implications of Native Tree Species and Associated Vegetation of Lobo, Batangas, Philippines."

In 2016, he presented his research paper in the 3$^{rd}$ International Conference on Agriculture and Forestry (ICOAF) held in Sofitel Manila, Philippines. The title of his research was "Leaf Anatomical Characteristics of Native Tree Species in the Philippines."

Prior to his appointment as an instructor in the university, he also worked as a research staff in the project entitled: "Ecological Assessment and Monitoring of Biodiversity in Terrestrial and Aquatic Ecosystems in Didipio Gold Copper Project, Nueva Vizcaya, Philippines."

**Honors:** When he presented his paper in INTECOL 2017, he was a recipient of the "Young Scholars Award" by the INTECOL and People's Republic of China.

In 2016, he was also a recipient of the 'Best Paper Presenter" award given during the ICOAF 2016 in Manila, Philippines.

For. Hernandez graduated '*Cum laude*" in 2016 at the UPLB.

**Recent Publication:**

Hernandez, J. O., P. L. Malabrigo Jr., M. O. Quimado, L. SJ. Maldia, & E. S. Fernando. 2016. Xerophytic characteristics of Tectona philippinensis Benth. & Hook. *Philippine Journal of Science* 145(3): 259–269.

## Pil Sun Park

**Affiliation:** Department of Forest Sciences, College of Agriculture and Life Sciences, Seoul National University

**Education:** PhD

**Research and Professional Experience:** *Research on Forest Ecology and Silviculture*

- Studying vegetation changes and forest stand structure related to disturbance and environment
- Integrating field information into practical applications in silviculture and forest restoration

**Professional Appointments:** Professor in Silviculture and Forest Ecology

## *Don Koo Lee*

**Affiliation:** Department of Forest Sciences, College of Agriculture and Life Sciences, Seoul National University

**Education:** Dr. Lee earned his Master's degree in forest genetics in 1971 from SNU and Master's degree in forest biometry in 1975 from Iowa State University, where he studied sampling techniques on estimating distribution and quantity of hybrid poplar roots. In 1978, he earned his PhD degree in silviculture from Iowa State University as well, where he made research on the influence of geometry and distribution of root systems on coppice regeneration and growth of hybrid poplars.

**Research and Professional Experience:** Dr. Don Koo Lee, from the Republic of Korea (ROK), is a renowned expert in forest sciences, especially silviculture and forest biometry. On 29 February 2012, Dr. Lee, Professor of silviculture and restoration ecology, retired as a faculty member of the Department of Forest Sciences, College of Agriculture and Life Sciences (CALS), Seoul National University (SNU) after 31 years of service. During his tenure, Dr. Lee distinguished himself in numerous ways and has contributed at a consistently high level to the department, college, and university. From 1981 until his retirement, he mentored 40 Master's and 20 PhD students, including foreign students from Asia. His academic career culminated in the publication of "Ecological Management of Forests," a book he authored with 29 of his students. He is currently a Professor Emeritus at SNU.

As a scientist, he worked on various research activities in tropical countries of Southeast Asia (including Cambodia, Indonesia, Lao PDR, Malaysia, Myanmar, Philippines, Thailand and Vietnam) through the ASEAN-Korea Environmental Cooperation Project (AKECOP) as Project Leader from 2000 up to 2011. AKECOP is an environmental collaboration initiative which facilitates forest cooperation on research and education between Korea and tropical countries of Southeast Asia under the theme "Restoration of Degraded Forest Ecosystems in Southeast Asian Tropical Regions."He also worked as Project Leader of R&D Center for Restoration of Forest Ecosystem Functions on Different Forest Zones (CERES) which focused on the restoration of degraded forests in three eco-regions, including tropical ecosystem from 2006 up to 2009. Furthermore, he became the Project Investigator of the "Cooperative Research for Restoration of Degraded Ecosystems in Northeast Asian Region" from 1999 up to 2008 with support from the Korea Science and Engineering Foundation (KOSEF).

**Professional Appointments:** He served as Dean of the College (1999 – 2001), Director of the National Instrumentation Center for Environmental Management (NICEM) (1995 – 1996), and Director of Agricultural Library (1997 – 1999). He served as Adjunct Professor in China, in particular at the Institute of Applied Ecology, Chinese Academy of Sciences (2005 – 2010) and Yanbian University (2005). He also became a Distinguished Guest Professor at Beijing Forestry University, China in 2007, and a Visiting Professor at Oregon State University, USA in 2009.

Dr. Lee became a member of the Scandinavian Plant Physiology Society (1985-2005), Korean Academy of Science and Technology (KAST) since 1994, Board of Trustees of the Center for International Forestry Research (CIFOR) (1999 – 2004), Society of American Foresters (2002 - 2012), Royal Swedish Academy of Agriculture and Forestry (KSLA) since 2003, Board of Directors of Seoul Green Trust (2006 – 2011), Editorial Board of the Chinese Forestry Science and Technology Journal (2006), Legislative Support Group of the National Assembly of the Republic of Korea (ROK) (2007), and Board of Directors of Climate Change Center of the Korea Green Foundation (2008 – 2010). He served as President of the Korea Forestry Energy Research Society (1998 – 2001), Korean Association of Societies for Agricultural Sciences (KASAS) (2000 – 2002) and Korean Forest Society (2004-2006), Chair of the Steering Committee of 'Forest for Peace' (1999 – 2006), and Co-representative of the Northeast Asian Forest Forum (NEAFF) and 'Forest for Life' (2003-2011 and 2004-2011, respectively).

He has served the International Union of Forest Research Organizations (IUFRO) from 1996 up to 2014 as Enlarged Board Member, Vice President, President and Immediate Past President, respectively. As officeholder of IUFRO for the past years, he has contributed to the promotion of global cooperation in forest-related research. During his term as President of IUFRO, the XXIII IUFRO World Congress was successfully convened in Seoul in 2010 with the Government of ROK as host country. Having this kind of international event with global influence and significance, impressive research developments as well as innovative sustainable forestry practices in ROK were showcased.

Dr. Lee has continuously gained his distinction when he was appointed by the President of ROK as Minister of the Korea Forest Service (KFS) in February 2011. It was the first time in the history of KFS that its Minister came from the academic sector. It was under his leadership that KFS obtained top rank in one-year accomplishments among 38 government organizations in ROK in 2011. Also, through his leadership, the Asian Forest Cooperation Organization (AFoCO) has been initiated in 2012 and established in as a legal entity of regional organization under an intergovernmental multilateral arrangement involving ASEAN Member States, ROK and other Asian countries since May 2018. In addition, under his term as Minister, the 10th Session of the Conference of the Parties of the United Nations Convention to Combat Desertification (UNCCD COP 10) was successfully held in ROK in October 2011. He proposed the Changwon

Initiative, which provides practical measures to battling desertification and land degradation especially African countries. As former President of the UNCCD COP 10, he made every effort in his capacity to work in close cooperation and full coordination with all country Parties and regional groups to respond with the different UNCCD challenges. He facilitated bilateral cooperation between KFS and governmental or international agencies of tropical countries (CATIE in Costa Rica, Ministry of Environment of Ecuador, Ministry of Environment of Brazil) since June 2012.

**Honors:** Dr. Lee has been the recipient of various awards, including the 1980 Scientific Achievement Award by the Korean Forest Society, First Gasan's Scientific Achievement Award by Korea Soho Culture Foundations in 2002, and 14th Evergreen Grand Award (Sangnok Daesang) by CALS, SNU in 2006. In July 2007, he received the Honorary Doctoral degree from Moscow State Forest University in Russia in recognition of his contribution to the field of forestry and to the development of global scientific forest network. In 2013, he received the Yellow Stripes Order of Service from Korea Government. In 2014, he received the Honorary Membership award from IUFRO and the George Washington Carver Distinguished Service award from Iowa State University, USA in October 2015.Last November 2017, he received the Honorary Doctoral degree from Kasetsart University in Thailand for his contribution to the restoration of degraded forest ecosystems in ASEAN region.

His dream for ROK is to continue to lead and collaborate with the international community in sustainable development and forestry cooperation, especially in developing countries.

During his academic career, Dr. Lee was able to produce around 350 publications, including articles in various scientific journals and books with his students written in Korean and English. The book series "Keep Asia Green" includes rehabilitation and restoration efforts in Asia, which employed a wide array of ecological, social and economic approaches. 'Volume I' provides information about the forest status and causes of degradation in eight tropical countries in Southeast Asia (Volume II in

Northeast Asia, Volume III in South Asia, and Volume IV in Central and West Asia). This publication aims to better understand the national capacities in terms of forest rehabilitation and existing education programs and analyze the need for further strengthening forest landscape restoration efforts in each country.

**Recent Publication:**
Lee, D.K., M.N. Salleh, W.M. Ho, and M.S. Combalicer. 2016. Changing Trends of Forestry Research Demand. Eds. Pancel, L. and M. Kohl. *Tropical Forestry Handbook*. Second Edition. Springer-Verlag Berlin Heidelberg. pp. 3559 -3570.

In: *Acacia*
Editor: Aide Matheson

ISBN: 978-1-53614-237-2
© 2018 Nova Science Publishers, Inc.

*Chapter 4*

# NMR Spectroscopy in Solution of Gum Exudates Located in Venezuela

## *Maritza Martínez, PhD*[*] *and Gladys León de Pinto, PhD*[±]

Centro de Investigaciones en Química de los Productos Naturales "Dra. Gladys León de Pinto," Facultad de Humanidades y Educación, Universidad del Zulia, Maracaibo, República Bolivariana de Venezuela

### Abstract

NMR spectroscopy applied to the study of gum exudates has become important since the 1990s in Venezuela. Analytical and structural studies of 23 .species, belonging to different genera and families have been reported, through the combination of classic methodology for carbohydrates and NMR spectroscopy. Up to the present, four types of heteropolysaccharides constituents of these gums have been described: A,B,C and C. Gums of the genera *Acacia*, *Albizia*, *Pithecellobium* and *Spondias*, among others, involve heteropolysaccharides of the first type. The core is represented by a ß (1 ---> 3) galactan, in some cases, with residual uronic acids and ß-L-arabinopyranose. These polysaccharides

---

[*] Corresponding Author: mmartinez.luz@gmail.com.
[±] Dedicated to her memory.

differ basically in the composition and sequences of the L-arabinose side chains. It was evidenced, by 2D-NMR correlation spectroscopy, a carbohydrate-protein linkage in the core of *Acacia tortuosa* gum. On the other hand, characteristic features of type B heteropolysaccharides were recorded, with some variations, for *Sterculia apetala*, *Cedrela odorata* and *Pereskia guamacho* gum exudates while *Cercidium praecox* heteropolysaccharide corresponds to a C structure, with a ß (1--->4) xylan core. These NMR techniques contributed to identify and confirm relevant structural features of the different gum exudates of species located in Venezuela.

**Keywords:** gum exudates, heteropolysaccharides, NMR spectroscopy, classic methodology for carbohydrates

## 1. GENERALITIES OF THE GUM EXUDATES

Gums are considered hydrocolloids capable of modifying the properties of the rheological systems in which they are involved (Rincón et al., 2014).

They can be classified, according to their origin, as gum exudates, seed gums, extracts proceeding from the bacterial ferment and modified gums (Carr, 1993)

Gum exudates are complex polymers excreted by tropical and subtropical regimes of the world (Jones and Smith, 1949). Exudation occurs when the plant is subject to stressing agents such as: mechanical wounds (practiced at trunk or branch level), the invasion by insects or fungi or the inoculation by chemical agents (etephon; methyl jasmonate) (Babu and Shah, 1987; Stephen et al., 1990). Dry environmental conditions favor gum production (Clamens et al., 2000); nevertheless, Stephen et al., 1990 describes that no specific climatic conditions are required for exudation to occur. Although the bulk of commercial supplies come from arid lands, any type of region, from rain forest to temperate and artic zones, may be a source.

Gum exudates, from the chemical point of view, are acid heteropolysaccharides made up of different types of aldoses: hexoses (galactose, glucose, mannose), pentoses (arabinose, xylose) and deoxyhexoses (rhamnose, fucose) The presence of glucose, as an atypical feature, was identified for the first time in *Anacardium occidentale* gum (Anderson et al., 1991a) and ribose has been detected in *Acacia senegal* gum (Singh et al., 2015). Acidity is represented by the following uronic acids: glucuronic, 4-O-methyl-D-glucuronic and galacturonic acids (Jones and Smith, 1949). It has been reported that this latter acid occurs in plant gums alone or interspread with rhamnose in the interior chains of the complex molecular structures (Anderson et al., 1991a).

The constituent heteropolysaccharide of the gum exudate can be structurally classified in four types: A-D (Stephen, 1983; 1990)

Group A, the most abundant, is conformed by a nucleus 3,6-di-O-D-galactan with units of L-arabinofuranose. The so-called type II AG (Aspinall Type II) has ramifications of other sugars: L-rhamnose, D-xylose and D-glucuronic acid (Aspinall, 1982).

The relatively uniform backbones of *Acacia* gums are made up of blocks of galactose (1--->3) of small size (molar mass 2000) separated by residues of 6-O- galactose (Stephen, 1983).

The residues present in the peripheral regions, surrounding the galactan, are relatively constant in *Acacia* gums and others type A heteropolysaccharides. The common linkages are: 4-O-OMe-α-D-GlcA (1- ->4)Gal, L-Rham$_p$ (1-->4)GlcA, L-Ara$_f$(1--->3)L-Ara$_f$. The last two are, labile to the acid medium (Stephen, 1983).

Enzimatic treatment has proved that carbohydrate chains can be attached as O-glycosides to hydroxyproline and/or serine and/or aspartic and/or glutamic acid (Stephen, 1983).

Group B includes those exudates in which the main nucleus is a galacturonan (1-->4) with alternate residues of 2-O-L-rhamnose. Karaya gum (*Sterculia urens*), an example of this type B heteropolysaccharide, is highly acetylated, with β-D-glucuronic acid and β-D-galactose terminal residues, attached to 0-2 and/or 0-3 of 4-0- α -D-galacturonic acid (Stephen, 1983).

Group C constitutes an atypical structure: a 4-O-xylan with short ramifications of L-arabinofuranose, galactose and rhamnopyranose.

Solubility in water improves when the substituents along with the 0-2 and 0-3 of the xylan are eliminated. The gum exudates of *Cercidium praecox* (León de Pinto et al. 1994), *Livistona chinensis* ("Chinese palm") and other Araceae are examples of this heteropolysaccharide group. (Maurer-Menestrina et al., 2003; Simas, 2004; 2006; Simas et al., 2014).

In the type D exudates, the mannuronic acid is present as a main nucleus. The structure of the ghatti gum (*Anogeissus latifolia*) represents this group (Stephen, 1983).

It should be noted that two gums, with the same type of nucleus, (for example an arabinogalactan type II) can present structural features which differentiate them. The ramifications can be made up of 3-O-α-L-arabinofuranose residues and 4-O-β-D-glucuronic acid, terminated in arabinofuranose, galactose, and rhamnose, as in the Arabic gum, or they can have 5-O-L-arabinofuranose residues and terminal β-D-glucuronic acid predominantly, as in the gum of *Meryta sinclarti* (Sims and Furneawax, 2003). Also, an exudate can include molecules of two types of heteropolysaccharides, i.e., A and B, as it occurs in the khaya gum (Sims and Furneawax, 2003).

The use of gel filtration and hydrophobic affinity chromatography has contributed to demonstrate that these exudates constitute heterogeneous systems in which an important protein fraction, associated to the carbohydrate, exists in the form of AGP complexes and glycoproteins (Randall et al., 1988). In addition to this, the presence of a lipidic anchor (GPI), linked to the AGP complex, has been described for some gum exudates. This last component is probably involved in the emulsifying processes (Yadav and Nothnagel, 2006).

The gum exudate of most commercial interest is the "Arabic gum", which is produced in the so-called "African gum- producing belt". The economical and political problems to which the region has been subjected since 1972, have incentivated the study of gum exudates of species of other genera and families. (Anderson et al., 1991b).

Advances in NMR spectroscopy techniques are stimulating renewed interest in their use for characterizing structural features in gum exudates or their degradation products (from oligosaccharide level) since 1980 (Defaye and Wong, 1986; Stephen et al., 1990)

Since 1990, León de Pinto G. and her research group, in the Centro de Investigaciones en Química de los Productos Naturales, Universidad del Zulia, Venezuela, have performed physicochemical and structural studies of the gum exudates of approximately 23 species, of diverse genera, combining the classic methodology for carbohydrates with 1D- and 2-D NMR techniques.

Researchers of other countries have also used these spectroscopic techniques as tools for the analytical and structural study of other gums. Anderson et al., (1991b) resolved the problem about the variations in the analytical parameters of different samples of *Combretum nigricans* gum using NMR spectroscopy. McIntyre et al., (1996) applied one- and two-dimensional NMR techniques to various heteropolysaccharides, including Arabic gum. Also, the application of the NMR spectroscopy, allowed the determination of the chemical structure, the degrees of branching and polymerization, and the conformation of the individual sugar units in oligosaccharides of this last gum (Tischer et al., 2002).

The assignment of the signals, sequences, and conformation of the polysaccharides can be done by the use of different methods. In our research, the indirect strategy was used, assuming that the chemical displacements of the monomers are similar to those of the equivalent carbons in the polysaccharide. The glycosidic linkage causes a displacement at low field (α effect) of the signals of the carbons involved, while those corresponding to the neighboring carbons experiment a displacement toward a high field (β effect) (León de Pinto and González de Troconis, 1989).

This review aims to provide a selected survey of the papers published in this area in the past 25 years, in the Centro de Investigaciones en Química de los Productos Naturales "Dra Gladys León de Pinto", Universidad del Zulia, Venezuela, about gum exudates proceeding from plant species highly disseminated in Venezuela.

## 2. Methodological Route Used in the Study of the Gum Exudates

### 2.1. Degradation of the Original Polymer

The purified polymer was degraded in two main ways, which allowed the gathering of information about the core and the ramfications: a) partial acid hydrolysis with $NaIO_4$ drastic oxidation and b) successive Smith degradations.

### 2.2. NMR Techniques Used

#### 2.2.1. 1D-$^1H$ NMR Spectroscopy

The NMR spectra of protons, of the original and degraded polymers, were not too useful for the structural elucidation of these complex materials, due to the overlapping of the resonances of the secondary protons of the rings (3.00-4.42 ppm). Nevertheless, they provided important information on the so-called "appendix groups": acetyls (-OAc, $-CH_3$, $-CH_2$-), at high field, and anomeric protons (between 4.4-5.5 ppm). For example, the C-1 of the α anomer resonated at lower field than the corresponding to the β anomer in the $^4C_1$ conformation of D-pyranose residues. Additionally, the anomeric region, allowed to indicate the number of linkages present in the polysaccharide.

#### 2.2.2. 1D-$^{13}C$ NMR Spectra

The polysaccharides, original or degraded, were dissolved in $D_2O$ in such a way that the –OH were fully deuterium exchanged. 1,4-dioxane (67.5 ppm) and occasionally methanol (49.5 ppm) were used as internal standards.

1D-Carbon-13 NMR (with broadband proton decoupling) provided information about the different monomers present in a polysaccharide, type of ring (furanose or pyranose), absolute configuration (D or L), anomers (α or β), and also about the glycosidic linkages (Agrawal, 1992). Frequently, there were good correlations between the chemical shifts of the polysaccharide and the chemical shifts of the component sugars (monomers). The chemical displacements obtained were compared to those obtained for analogous polymers (Bock and Pedersen, 1983; Utille et al., 1986; Tischer et al., 2002).

### 2.2.3. DEPT-135

The use of the DEPT-135 technique, in comparison to the 1D-Carbon-13 NMR spectrum, permitted to assign, in an unequivocal way, the -OCH$_3$ y -CH$_3$ (non-inverted signals) and -CH$_2$-, (inverted signals) of hexoses and pentofuranoses present in the polymer (Martínez et al., 2015).

### 2.2.4. Two-Dimensional (2D) Correlation NMR Spectroscopy

The applications of the two-dimensional (2D) correlation NMR for the structural elucidation of the studied gum exudates are described in Table 1.

**Table 1. Usefulness of the two-dimensional (2D) correlation NMR**

| Technique | Description<br>The technique correlates: |
|---|---|
| COSY | Adjacent protons $^1$H shifts in a ring residue |
| TOCSY | Almost all $^1$H resonances within a monosaccharide residue |
| HMQC, HSQC | The $^{13}$C and $^1$H shifts of directly bonded C–H |
| HMBC | The $^{13}$C and $^1$H shifts through scalar long-range linkages (up to three) |

## 3. INTERPRETATION OF THE SPECTRA

The interpretation of 1D-$^1$H and $^{13}$C-NMR spectra, of the original polysaccharide and of its degraded products, was done in crescent order of its chemical complexity. The assignments performed on the spectra of the simpler polymers served as a basis to interpret the spectra of the more complex ones.

On the other hand, the information collected by 1D-, served as the basis for the interpretation of the 2D- spectra.

The assignment of the signals in the spectrum of the simpler polymer (the nucleus of the structure), was done by contrasting with chemical displacements of adequate model compounds, reported for monomers, oligomers, and analogous polymers (Bock and Pedersen, 1983; Utille et al., 1986; Joao et al., 1988).

The isolation of oligosaccharides, by partial acid hydrolysis of the polymer, in order to obtain simpler, easier, and more precise NMR spectra, whose signals by transitivity were applied to the most complex polymers, was also very useful. Structural features of the gums from *Anadenanthera colubrina* and *Acacia senegal* have been characterized by isolating and analyzing the structure of oligosaccharides which are usually present in the gum (Delgobo et al., 1999; Tischer et al., 2002).

For the interpretation of the Carbon-13 spectra, it was also useful to take into consideration the taxonomic origin (family, genus, amongst others) of the gum. The relationship between the molecular structures of gum exudates and their taxa of the origin has been reviewed (Stephen et al., 1990).

## 4. RELEVANT SPECTROSCOPIC FEATURES OF THE GUM EXUDATES FROM THE MAIN PRODUCING PLANT GENERA DISSEMINATED IN VENEZUELA

### 4.1. Type A Heteropolysaccharides

#### *4.1.1. Acacia Gum Exudates*

##### 4.1.1.1. *Acacia macracantha* and *Acacia tortuosa* Gums

*Acacia macracantha* (syn *Vachellia macracantha*) and *Acacia tortuosa* (syn *Vachellia tortuosa*) polysaccharide gums correspond to type II AGs (Aspinall Type II), with a gal:ara proportion of 2:1 and 3:1, respectively (Martínez et al., 1996a; 2015; León de Pinto et al., 1998a).

The gum of *A. tortuosa* has xylose, as an atypical feature, and it does not contain rhamnose. The presence of xylose was corroborated by degradation studies.

The nucleus of both gums, according to the chemical data, corresponds to a β-D-galactan, with residues of uronic acids and L-arabinose, resistant to the oxidation process with metaperiodate.

Spectroscopic evidences confirm these results obtained by chemical methods. The spectrum of the degraded B gum of *A. macracantha*, Figure 1, for example, showed an intense signal due to the C-6 linked of β-D-galactose, in two environments (68.4; 68,6 ppm, possibly correlated with the resonance, which appeared at low field (175 ppm), assignable to carboxylic groups of uronic acids (Martínez et al., 1996a).

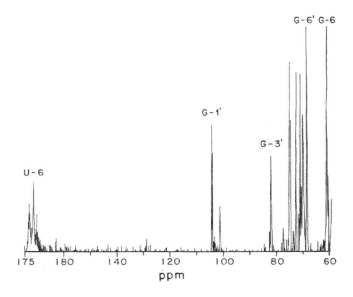

(Reprinted from *Carbohydrate Polymers*, 29, Martínez M., León de Pinto G., Rivas C. and Ocando E., Chemical and spectroscopic studies of the gum polysaccharide from *Acacia macracantha*, 247-252, Copyright 1996 with permission from Elsevier).

Figure 1. $^{13}$C-NMR spectrum of degraded gum B of *A. macracantha* gum. G = β-D-galactose. ' = carbon atom involved in the glycosidic linkage. U-6 = C-6 of uronic acids.

These uronic acid residues are vulnerable to the oxidation with metaperiodate, but possibly, steric hindrance could prevent the formation of the corresponding complex. Based on the chemical and spectroscopic information data, a model was proposed, Figure 2, for an important complex fragment isolated from *A. macracantha* backbone structure.

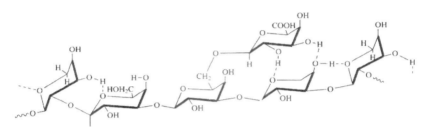

(Reprinted from *Carbohydrate Polymers*, 29, Martínez M., León de Pinto G., Rivas C. and Ocando E., Chemical and spectroscopic studies of the gum polysaccharide from *Acacia macracantha*, 247-252, Copyright 1996 with permission from Elsevier).

Figure 2. Suggested conformation of the structural fragment in the basis of methylation analysis and NMR assignment.

The 1D-Carbon-13 NMR spectroscopy allowed the detection of differences between *A. macracantha* and *A. tortuosa* gum exudates. For example, in the first one, there were identified 2-O-β-L-arabinopyanose side chains, finished in L-arabinofuranose, whereas in the second one, 3-O- L-arabinofuranose and/or 3-O- L-arabinopyranose chains, ended in L-arabinofuranose, were more evident.

On the other hand, studies undertaken by fractioning, demonstrated that the gum exudate of *A. tortuosa* is a heterogeneous system, with a high polydispersity index (Ð = 2), conformed by four fractions, one of which is an AGP complex (Beltrán et al., 2005).

This chemical information led to a deeper study of this exudate in relation to the nature of the carbohydrate-protein linkage. In this sense, the use of 2D- correlation spectroscopy techniques, homo- and heteronuclear, were very useful. (Martínez et al., 2015). The COSY spectrum of *Acacia tortuosa* degraded B gum, for example, showed a group of cross-signals characteristic to α/β aa, at low/high fields, between 4,1 - 4.9/1.1 - 1.9 ppm, Figure 3.

NMR Spectroscopy in the Solution of Gum Exudates Located ...    123

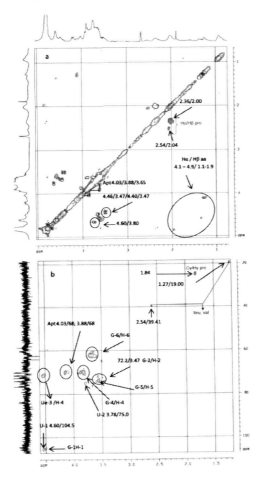

(Reprinted from *Food Chemistry*, 182, Martínez M, Beltrán O, Rincón F, León de Pinto G e Igartuburu JM, New structural features of *Acacia tortuosa* gum exudate, 105-110, Copyrightt 2015 with permission from Elsevier).

Figure 3. NMR spectral data for *Acacia tortuosa* degraded gum B (a) COSY (b) HMQC. Ap$_t$ = terminal arabynopyranose residues, U$_e$ = 4-0-Methyl-α-D-glucuronic acid, G = galactose residues. Hα/Hβ = protons α and β aminoacids. pro = proline, leu = leucine, val = valine, C$^γ$/H$^γ$ = proline carbon and proton $^γ$.

The most important spectroscopic finding, for the first time reported for Venezuelan gums, was found in the HMBC spectrum of this same polymer, with an outstanding correlation signal of 1.83/75.10 ppm, which suggested the presence of a galactose-hydroxyproline linkage, an amino acid abundant in its composition (Martínez et al., 2015). Figure 4.

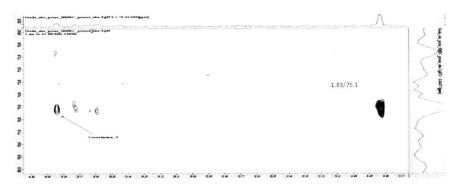

(Reprinted from *Food Chemistry*, 182, Martínez M, Beltrán O, Rincón F, León de Pinto G e Igartuburu JM, New structural features of *Acacia tortuosa* gum exudate, 105-110, Copyright 2015 with permission from Elsevier).

Figure 4. HMBC spectrum from *Acacia tortuosa* degraded gum B.

The two-dimensional spectroscopy of *A. tortuosa* original gum also led to a more exhaustive study about the structural features of its polysaccharide. It was corroborated the presence of:

(a) 3-O- β-D- galactose in four different environments.
(b) β-L-arabinopyranose
(c) 4-O-methyl-α-D-glucuronic acid as has been reported for the mezquita gum (López-Franco et al., 2006).

**4.1.1.2. *Acacia glomerosa* Gum**

On the other hand, *Acacia glomerosa* (syn *Senegalia polyphylla*), another species of the same genus, exudes a clear gum which forms a gel stable at environmental temperature, not requiring the presence of cations. The exudate has a high content of rhamnose (15%) similar to that reported for the gum of *A. senegal* (León de Pinto et al., 2001a).

As in *Acacia tortuosa* gum described above, it was demonstrated by SEC (open column), that the proteic fraction is involved in an AGP complex and also in low molecular weight peptides (León de Pinto et al., 2002a)

The 1D-¹H-NMR spectroscopy allowed the spectroscopic characterization of an oligosaccharide, which was isolated in the majority of its structure, conformed by galactose, rhamnose and glucuronic acid, Figure 5.

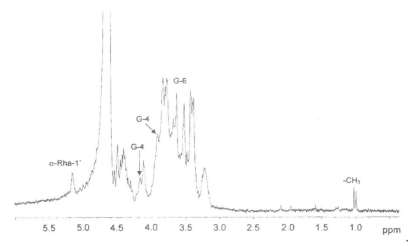

(Reprinted from *Food Hydrocolloids*, 15, León de Pinto G., Martínez M. and Sanabria L, Structural features of the polysaccharide gum from *Acacia glomerosa*, 461-467, Copyright 2001, with permission from Elsevier).

Figure 5. ¹H-NMR spectrum of the oligosaccharide isolated during the preparation of degraded gum A from *Acacia glomerosa* gum. G = β-D-galactopyranose, A_f = α-L-arabinofuranose, U = β-D-glucuronic acid, U_e = 4-0-methyl-α-D-glucuronic acid,'= carbon involved in the glycosidic linkage.

The degradation studies demonstrated that the core is also a β-(1-->3) galactan, as well as the *Acacia* gums described above. The two-dimensional spectroscopy of the polysaccharide II, almost the nucleus of the structure, Figure 6. corroborated the existence of two environments for 3-O-β-D-galactose and the persistence of uronic acids (β-D-glucuronic and its 4-OMe-α-ether), probably linked to the C-6 of galactose, as was described for *A. macracantha* and *A. tortuosa* gums

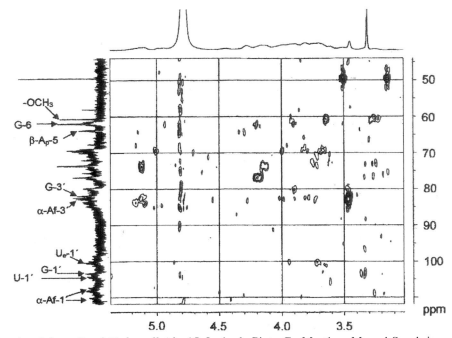

(Reprinted from *Food Hydrocolloids*, 15, León de Pinto G., Martínez M. and Sanabria L, Structural features of the polysaccharide gum from *Acacia glomerosa*, 461-467, Copyright 2001, with permission from Elsevier).

Figure 6. Bidimensional spectroscopy studies of polysaccharide II from *A. glomerosa* gum.

The spectral data indicated that, even in this polymer II, L-arabinose residues (furanose and pyranose) are present No resonances were observed that corresponded to the proteic fraction involved with the carbohydrate, as in the case of *A. tortuosa* gum.

### 4.1.2. Albizia Gum Exudates

#### 4.1.2.1. *Albizia lebbeck* Gum

The *A. lebbeck* ("lara") gum, as the *A. tortuosa* gum exudate, does not have rhamnose, as was confirmed by the absence of the methyl $-CH_3$ signal in the spectrum of the original polysaccharide, in correspondence with its chemical composition (Martínez et al., 1996b).

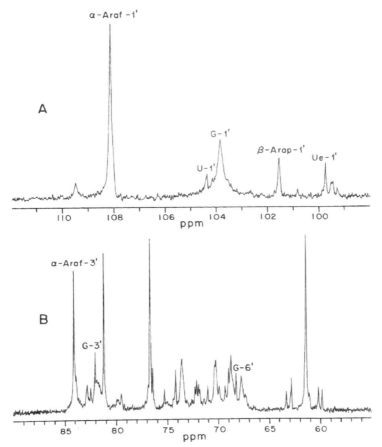

(Reprinted from *Biochemical Systematics and Ecology*, 23, Martínez M., León de Pinto G., Alvárez S., González de Troconis N., Ocando E and Rivas C., Composition and properties of *Albizia lebbeck* gum exudate, 843-848, Copyright 1995, with permission from Elsevier).

Figure 7. Expansion of [13]C-NMR spectrum of *Albizia lebbeck* gum in the range 95-110 PPM (2A) and in the range 60-65 PPM (2B). Ara$_f$ = arabinofuranose, Ara$_p$ = arabinopyranose, G = β-D-galactopyranose, U$_e$ = 4-0-methyl-α-D-glucuronic acid and U = β-D-glucuronic acid.

The signals of galactose overlapped with those of mannose (another hexose), present in the structure. Noteworthy is the resonance, at 79.54 ppm, in the region of the glycosidic linkages, assignable to C-3 of β -D-galactose residues linked to L-arabinose and/or assignable to C-4 linked β-D-galactose. This signal was not detected for *Acacia* gums. There were also evidences of 3-O-α-L-arabinofuranose side-chains, as was described for *A. tortuosa* gum, terminated probably in β-L-arabinopyranose, Figures 7A and 7B. The successive Smith degradation studies demonstrated that the length of these chains is up to two residues (León de Pinto and González de Troconis, 1989).

The signals of the nucleus, a β-(1-->3) galactan, were well defined for all the carbons belonging to 3-O- β-D-galactose. There were no chemical or spectroscopic evidences of residual uronic acids and arabinose, as in the case of the *Acacia* gums. (León de Pinto and González de Troconis, 1989).

The signals of the nucleus, a β-(1-->3) galactan, were well defined for all the carbons belonging to 3-O- β-D-galactose. There were no chemical or spectroscopic evidences of residual uronic acids and arabinose, as in the case of the *Acacia* gums. (León de Pinto and González de Troconis, 1989).

### 4.1.2.2. *Albizia niopoides* var colombiana Gum

The gum of *Albizia niopoides* var colombiana, unlike the exudate of *A. lebbeck*, has a high content of rhamnose (21 - 24%), which was evident in the intense signal that appeared in the 1D-[13]CNMR spectrum of the original gum, at 17.42 ppm, Figure 8 (León de Pinto et al., 2002b).

The high content of rhamnose is comparable to that reported for *A. forbessi* and higher than that described for gums of other *Albizia* (Anderson and Morrison, 1990). Mannose was not detected. The presence of this sugar is of great diagnostic for *Albizia* gums but *A. niopoides* var colombiana deviates from this feature (León de Pinto et al., 2002b)

Unlike the spectrum of *A. lebbeck* gum, it was observed a resonance, at a very low field (181 ppm), assignable to α-D-glucuronic acid residues (substituted by metals). This spectroscopic finding was in correspondence with its high mineral content (19%)) (León de Pinto et al., 2002b).

In addition to this, it was detected the same signal, at 79.86 ppm, due to C-3 of galactose linked to L-arabinose and/or assignable to C-4 linked β-D-galactose, described for *A. lebbeck* gum.

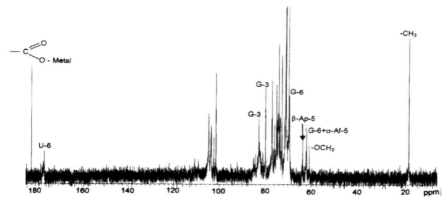

(Reprinted from *Ciencia*, 10, León de Pinto G., Martínez M., Beltrán O., Rincón F., Clamens C., Igartuburu JM, Guerrero R. and Vera A., Characterization of polysaccharides isolated from gums of two venezuelan specimens of *Albizia niopoides var. colombiana*, 382-387, Copyright 2002, with permission from Ciencia).

Figure 8. $^{13}$C-NMRspectrum of *A. niopoides* var. colombianagumexudate (0 – 200 ppm). G = β-D-galactose U= Uronic acids. ´= Carbon involved in glycosidic linkage. –OCH$_3$ = methoxyl group of 4-O-Me-α-D-glucuronic acid. -CH$_3$ = methyl group of rhamnose.

### 4.1.3. Pithecellobium Gum Exudates

#### 4.1.3.1. *Pitthecellobium saman* Gum

The *P. saman* (syn *Samanea saman*) gum exudate has structural features analogous to the *Acacia* gums, nevertheless it is less soluble in water, perhaps on account of its acetyl content, and it is more viscous (75 mL/g) (León de Pinto et al., 1995).

Two oligosaccharides were isolated from its structure, which demonstrated the presence of:

β-D-glucuronic acid $^1$-----> $^6$ β-D-galactose

α-D-glucuronic acid $^1$-----> $^4$ β-D-galactose

The degraded gum B, nucleus of the structure, is a β-(1-->3) galactan with residual uronic acids not vulnerable to metaperiodate oxidation. The spectroscopic study of this polymer allowed the visualization of the signals, not only of 3-O-β-D- galactose but also of 6-O-β-D- galactose, probably linked to the uronic acids (β-D-glucuronic acid and its 4-OMe-α-ether). Residual arabinose was not detected unlike it was observed for the gums of *Acacia*. (León de Pinto et al., 1998b).

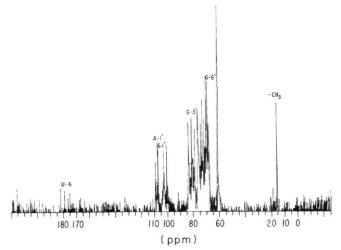

(Reprinted from *Biochemical Systematics and Ecology*, 23, León de Pinto G., Martínez M., Beltrán O., Gutiérrez de Gotera O., Vera A., Rivas C. and Ocando E. Comparison of Two *Pithecellobium* Gum Exudates, 849-853, Copyright 1996, with permission from Elsevier).

Figure 9. [13]C-NMR spectrum of saman gum. U = Uronic acids A = α-L-arabinofuranose G = β-D-galactose. ´ = Carbon involved in a glycosidic linkage.

The spectrum of the original *P. saman* gum additionally revealed the presence of α-D-glucuronic acid (substituted by metals), probably linked to the C-4 of β-D-galactose, according to the chemical evidences, Figure 9, i.e., C-1 (98.33 ppm) and C-6 (181.17 ppm), as was described for *A. niopoides* var. colombiana gum.

The spectrum of the original *P. saman* gum also showed that the chains of arabinose are made up of 3-O-α-L-arabinofuranose. There were no evidences of L-arabinopyranose residues. It was observed an intense signal due to the –CH$_3$ of rhamnose (16.50 ppm), in correspondence with its high content (11%). These residues were removed during the preparation of the polysaccharide I, confirming its terminal character. In this spectrum, was also detected a signal at high field, of low intensity, assignable to acetyl - CH$_3$ groups, (19.94 ppm) (León de Pinto et al., 1998b).

### 4.1.3.2. *Pithecellobium mangense* (Syn *Chloroleucon mangense*) Gum

The 1D-C$^{13}$ NMR spectrum of *Pithecellobium mangense* (syn *Chloroleucon mangense*) gum, Figure 10, corroborated the absence of rhamnose, in contrast with what was described for the exudate of *P. saman*.

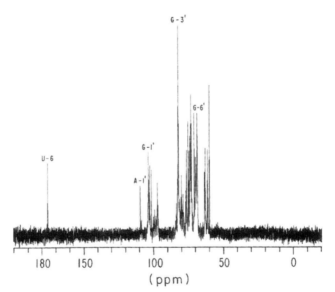

(Reprinted from *Biochemical Systematics and Ecology*, 23, León de Pinto G., Martínez M., Beltrán O., Gutiérrez de Gotera O., Vera A., Rivas C. and Ocando E. Comparison of Two *Pithecellobium* Gum Exudates, 849-853, Copyright 1996, with permission from Elsevier).

Figure 10. $^{13}$C-NMR spectrum of *P. mangense* GUM. U = Uronic acids A = α-L-arabinofuranose G = β-D-galactose. ´= carbon involved in a glycosidic linkage.

Nevertheless, this spectrum made evident the existence of 3-O-α-L-arabinofuranose chains, up to a maximum of four residues, in a form analogous to *P. saman* gum, but probably ending in β-L-arabinopyranose residues. Outstanding was the signal at 79.86 ppm, attributed to C-3 of β-D-galactose linked to L-arabinose, assignment supported by the fact that it disappears, by dearabinosilation, during the preparation of the degraded A gum (León de Pinto et al., 2001b). The absence of acetyl groups was also evidenced in this *P. mangense* spectrum, in contrast with the described for the gum exudate of *P. saman*. (Figure 10).

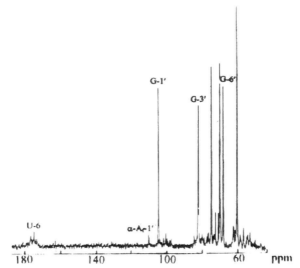

(Reprinted from *Carbohydrate Polymers*, 46, G.L. León de Pinto., M. Martínez M., E. Ocando and C. Rivas Relevant structural features of the polysaccharide from *Pithecellobium mangense* gum exudate, 261-266, Copyright 2001, with permission from Elsevier).

Figure 11. $^{13}$C-NMR spectra of the degraded gum B from *P. mangense* G = β-D-galactose U = Uronic acids, A$_f$= α-L-arabinofuranose ´= carbon involved in the glycosidic linkage.

The spectrum of the degraded B gum, Figure. 11, nucleus of the structure, showed that it is a β-(1-->3) galactan, with uronic acid (β-D-glucuronic acid and its 4-OMe-α-ether) and α-L-arabinofuranose residues, resistant to the metaperiodate oxidation, unlike the gum of *P. saman*. These residues are probably linked to the C-6 of 3-O-β-D-galactose, as was described for the *Acacia* gums. Outstanding was the signal of the C-6 linked β-D-galactose in two chemical environments (68.33 and 68.51 ppm).

### 4.1.4. Spondias Gum Exudates

#### 4.1.4.1. *Spondias purpurea* Gum

The gum of *S. purpurea* ("ciruelo de huesito" "bone plum"), contains mannose, xylose, and rhamnose, in addition to galactose and arabinose; the existence of galacturonic acid was chromatographically made evident (León de Pinto et al., 1996a). The presence of mannose in the exudates of these species is a general feature and could be associated to the existence of an AGP-GPI complex (arabinogalactan-glycosilfosfatidilinositol). It has been described that the GPI lipidic anchor might be responsible for the emulsionation processes in these exudates (Yadav and Nothangel, 2006).

The following aldobiouronic acids were isolated by partial acid hydrolisis of the original gum and its degraded products:

β-D-glucuronic acid$^1$------>$^6$galactose

4-O-methyl-α-D-glucuronic acid$^1$------>$^4$galactose

β-D-glucuronic acid$^1$------>$^6$galactose$^1$------>$^6$galactose

The nucleus of the structure is likewise a β(1-->3) galactan, as in the case of the polymers above described. By means of partial acid hydrolysis of the degraded gum B, a neutral oligosaccharide galactose$^1$------>$^3$galactose was isolated, as an indicator of the existence of blocks of this type of residues in the nucleus, unlike suggested for the rest of the gums studied.

The spectrum of the polysaccharide II, Figure 12, showed the signals characteristic to a β- (1-->3) galactan. No resonances were observed for C-6 linked β-D-galactose or for residual arabinose and uronic acids, as it was described for the gums of *Acacia*.

The spectrum of the original gum exhibited, in addition to the resonances for 3-O-β-D-galactose, signals for 6-O-β-D-galactose (C-6: 69.56 ppm). Likewise was identified the one corresponding to residues of 3-O-β-D- galactose linked to residues of arabinose (at 79.81 ppm), visualized in the spectra of *Albizia* gums. This signal was associated with the corresponding to the C-1 of terminal α-L-arabinofuranose (109.30 ppm) since it disappeared, by dearabinosilation, during the preparation of the degraded A gum (León de Pinto et al., 1998).

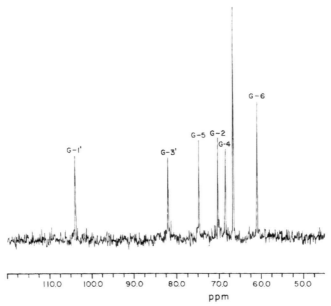

(Reprinted from *Carbohydrate Research*, 290, Gladys León de Pinto., Maritza Martínez, Julián Alberto Mendoza, Dinorah Ávila, Edgar Ocando and Carlos Rivas (Structural study of the polysaccharide from *Spondias purpurea* gum exudate, 97-103, Copyright 1996, with permission from Elsevier).

Figure 12. $^{13}$C-NMR spectrum of polysaccharide II from *S. purpurea* gum.
G = β-D-galactose.

The two-dimensional spectroscopy of the original gum, by combination of COSY, HMQC and HMBC tecchniques, allowed to deepen in the structural features for a sequence of L-arabinose residues: α-L-araf[1]------->[2] α-L-arap[1]--------->[3] α-L-araf (Martínez et al., 2008).

The two-dimensional spectroscopic study of *S. purpurea* polysaccharide II, Figures 13 and 14, allowed to make an approximation to the peptide associated to the nucleus of the structure. A characteristic resonance of hydroxyproline, abundant in this degraded polymer, was identified (In HMQC: Hσ/Cσ: 4.13;4.17/68.20 ppm), but there were also observed signals due to proline and leucine, probably more exposed to the field. There were no evidences in relation to the carbohydrate-peptide linkage as was described for the gum of *A. tortuosa* (Martínez et al., 2015).

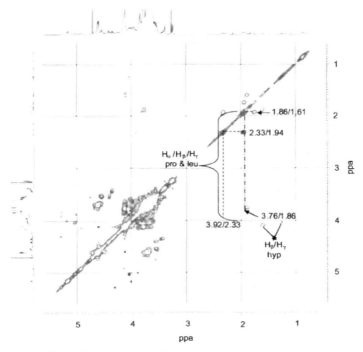

(Reprinted from *Food Hydrocolloids*, 22, M. Martínez, G. León de Pinto, M. Bozo de González, J. Herrera, H. Oulyadi, and L. Guilhaudis New structural features of *Spondias pupurea* gum exudate, 1310-1314, Copyright 2008, with permission from Elsevier).

Figure 13. COSY spectrum (1-5 ppm) of polysaccharide II from *S. purpurea*.

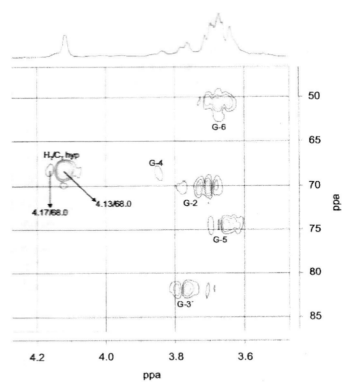

(Reprinted from *Food Hydrocolloids*, 22, M. Martínez, G. León de Pinto, M. Bozo de González, J. Herrera, H. Oulyadi, and L. Guilhaudis New structural features of *Spondias pupurea* gum exudate, 1310-1314, Copyright 2008, with permission from Elsevier).

Figure 14. HMQC spectrum of polysaccharide II from *S. purpurea*.

The peptide causes different chemical environments to be observed for the β- (1 -> 3) galactan. The absence of residual uronic acids and arabinose in the nucleus was corroborated.

**4.1.4.2. *Spondias dulcis* Gum**

The sugar composition of the gum of *S. dulcys* (syn. *Spondias cytherea*) "jobo de la India", is similar to the gum of *S. purpurea*, even though its content of rhamnose is higher (17%). It was demonstrated, by means of partial acid hydrolysis, that the residues of arabinose, xylose, rhamnose, and mannose are terminal in the structure (Martínez et al., 2003).

The degraded B gum corresponds to a β- (1 -> 3) galactan, with residual uronic acids, unlike the gum of *S. purpurea*. The two-dimensional spectroscopy, HMQC, of the degraded B gum, Figure 15, confirmed the existence of these acid residues in the nucleus, probably linked to the C-6 of residues of β-D-galactose. In addition to the cross signals for 3-O-β-D-galactose, resonances were unequivocally identified for 6-O-β-D-galactose (H-6/C-6-linked = 3.87/68.50 ppm); β-D-glucuronic acid (H-3/C-3 = 3.54/76.96 ppm); 4-O-methyl-α-D-glucuronic acid (H-4/-OCH$_3$ = 3.33/60.14 ppm; H-4/C-4 = 3.12/82.72 ppm).

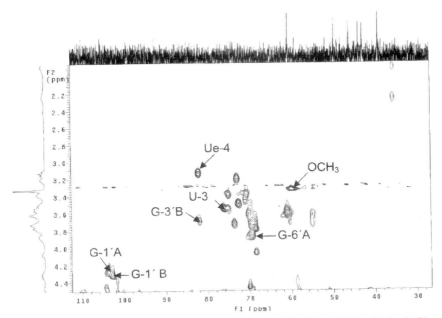

(Reprinted from *Carbohydrate Research*, 338, Maritza Martínez, Gladys León de Pinto, Lilian Sanabria, Olga Beltrán, José M. Igartuburu, Ali Bahsas, Structural features of an arabinogalactan gum exudates from *Spondias dulsis* (Anacardiaceae), 619-624, Copyright 2003, with permission from Elsevier).

Figure 15. HMQC of degraded gum B, obtained by oxidation of degraded gum A from *S. dulsis*, A = 3-O-β-D-galactose, B = 6-O--β-D-galactose, U = β-D-glucuronic acid.

No signals were identified for a peptide associated to the nucleus of the carbohydrate as for the gum of *S. purpurea*.

The two-dimensional spectroscopy of the original and degraded A gums confirmed the assignments done for 3-O- and 6-O-β-D-galactose and the uronic acids. It was also corroborated the presence of chains constituted by residues of 3-O-β-L-arabinopyranose: H-1/C-1: 5.2/100.68 ppm; H-3/C-3: 4.13/74.08 ppm; H-5/C-5: 3.45/63.40 ppm and/or 3-O-α-arabinofuranose: H-1/C-1: 5.2/108.60 ppm; H-3/C-3: 4.11/82.27; H-4/C-4: 4.17/82.80 ppm. The units of L-arabinopyranose are in β anomeric form, unlike the gum of *S. purpurea* (Martínez et al., 2003).

### 4.1.4.3. *Spondias mombin* Gum

The gum of *S. mombin* ("jobito"), in contrast to the rest of *Spondias* gums, does not have xylose (León de Pinto et al., 2000).

The metylation analysis revealed that it is a complex polysaccharide constituted by several types of glycosidic linkages. The L-arabinose exists predominantly in the furanose form, 2-O- and 3-O- linked. There are also chains of 2-O-L-arabinopyranose, but in a lesser proportion, finished in L-arabinofuranose. The degraded B gum is a β- (1 -> 3) galactan with residues of residual uronic acids, as for the gum of *S. dulcis*.

Alcaline degradation studies of *S. mombin* gum exudate, were carried out, taking into consideration its relatively high protein content. (León de Pinto et al., 2000). The elution profiles obtained by size exclusion chromatography (SEC), using scattered light and refraction index detectors, demonstrated that the gum of *S. mombin* is a heterogeneous system with a high degree of polydispersity (Đ = 3.8), with three populations of different molar masses.

The HSQC NMR spectroscopy constituted a valuable tool in the diagnosis of the efficiency of the alkaline conditions used. There were observed correlations, which made evident the persistence of different residues of aromatic and alifatic amino acids, linked by peptide bonds, in the hydrolyzed gum, Figure 16. Additionally, the absence of signals at ~180 ppm, assigned to carbonyl groups of free amino acids, allowed to suggest that a total rupture of the peptide did not occur (Leal et al., 2010). The HSQC

# NMR Spectroscopy in the Solution of Gum Exudates Located ... 139

showed some signals assignable to the Cδ of lisine and Cβ of the glutamic and aspartic acids, predominant in the structure, in accordance with the composition of amino acids determined by HPLC.

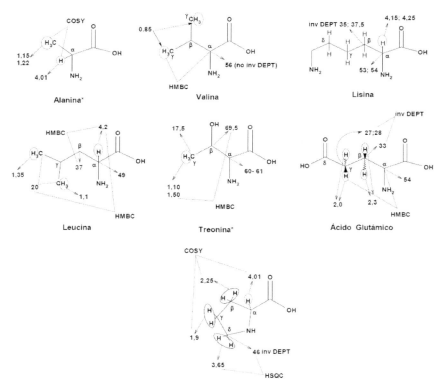

(Reprinted from *Revista Iberoamericana de Polímeros*, 11, María Leal, Marítza Martínez, Olga Beltrán, Gladys León de Pinto, Luc Picton, Didier Le Cerf, Hassan Oulyadi y Laure Guilhaudis Estudio de la interacción carbohidrato-proteina en goma de *Spondias mombin* por cromatografía de exclusión por tamaño (SEC) y Resonancia Magnética Nuclear (RMN), [Carbohydrate-protein interaction studies of *Spondias mombin* gum by size exclusion chromatography (SEC) and Nuclear Magnetic Resonance (NMR)], 495-510, Copyright 2010, with permission from Revista Iberoamericana de Polímeros).

Figure 16. Desplazamientos químicos (δ, ppm) de los aminoácidos presentes en la goma de *Spondias mombin*, después de la hidrólisis alcalina [Chemical shifts of the amino acids present in *Spondias mombin* gum].

The COSY, on the other hand, exhibited Hα/Hβ correlations attributed to the different residues of amino acids, Figure 16.

The $^{13}$C-NMR spectrum of the polysaccharide III, Figure 17, an approximation to the nucleus, showed signals for 6-O-β-D- galactose (C-1: 103.82; C-2: 74.53; C-3: 76.30; C-5: 73.09 and C-6: 68.52 ppm) β-D-glucuronic acid (C-1: 103.95; C.2: 70.27; C-3: 72.58; C-5: 76.23 and C-6: 175.07 ppm) and 4-O-methyl-α-D-glucuronic acid (C-1: 99.89; C-3: 73.09; C-4: 81.79; C-5: 70.87 -OCH$_3$: 60.94 ppm), confirming chemical evidences. No signals for L-arabinose were observed

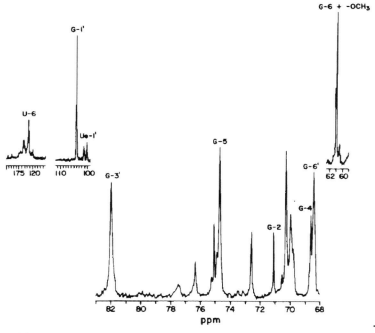

(Reprinted from *Carbohydrate Polymers*, 43, G. León de Pinto, M. Martínez, O. Beltrán, F. Rincón, J. Manuel Igartuburu and F. Rodríguez Luis. Structural investigation of the polysaccharide of *Spondias mombin* gum, 105-112, Copyright 2000, with permission from Elsevier).

Figure 17. $^{13}$C-NMR spectrum of polysaccharide III of *S. mombin*. G = β-D-galactopyranose, U = Uronic acids, U$_e$ = 4-O-methyl-α-D-glucuronic acid, A$_f$ = arabinofuranose, A$_p$ = β-L-arabinopyranose,' = carbón involved in the glycosidic linkage.

The $^{13}$C-NMR spectrum of the degraded A gum is much more complex than the spectrum of the polisaccharide III, Figure 18.

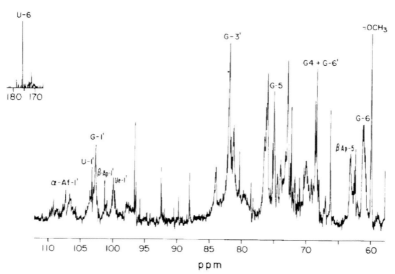

(Reprinted from *Carbohydrate Polymers*, 43, G. León de Pinto, M. Martínez, O. Beltrán, F. Rincón, J. Manuel Igartuburu and F. Rodríguez Luis. Structural investigation of the polysaccharide of *Spondias mombin* gum, 105-112, Copyright 2000, with permission from Elsevier).

Figure 18. $^{13}$C-NMR spectrum of degraded gum A of *S. mombin*. G = β-D-galactopyranose, U = Uronic acids, Ue = 4-O-methyl-α-D-glucuronic acid, $A_f$ = α-L-arabinofuranose, $A_p$ = β-L-arabinopyranose,' = carbón involved in the glycosidic linkage.

In addition to the signals for 3-O-β-D-galactose, this spectrum showed three chemical environments for 6-O-β-D-galactose (C-6 linked: 68.22; 68.34; 68.4 ppm), that allowed us to suggest that the C-6 constitutes an important branch point. Also, there were identified resonances for 3-O-α-L- and terminal α-L-arabinofuranose and 2-O-β-L- and terminal β-L-arabinopyranose, in accordance with the methylation analysis, which disappeared in the spectrum of the Polysaccharide III.

Additionally, two resonances were detected (at 54.83; 57.13 ppm) (region not shown) assignable to the carbon α of alifatic amino acids residues, in accordance with its high protein content (20.65%).

## 4.2. Type B Heteropolysaccharides

### 4.2.1. Cedrela Gum Exudates

#### 4.2.1.1. *Cedrela odorata* Gum

The nucleus of *Cedrela odorata* gum ("cedro") can be considered a variant of this type of heteropolysaccharides, because it is an acidic rhamnogalactan instead of a rhamnogalacturonan (Stephen et al., 1990). The presence of galactose has been described for the gum of *Khaya ivorensis*, another Meliaceae (Aspinall, 1970). D-galacturonic acid, a typical feature of these heteropolysaccharides, was not chemically identified (León de Pinto et al., 1996b).

The 1D-$^{13}$C-NMR spectrum of the original gum showed the intense signal of the –CH$_3$ de Ram (16.67 ppm). There were evidences for 3-O-β-D-galactose (C-1: 103.47; 104.15 ppm and C-3: 83.97 ppm), 3-O-α-L-arabinofuranose (C-1 107.57 ppm; C-3 83.97 ppm), and of 4-O-methyl-α-D-glucuronic acid (C-1: 99.31 ppm). There were no assignable resonances for 6-O-β-D-galactose (González de Troconis et al., 2001)

The spectrum of the Polysaccharide I also allowed to resolve the resonances in the carbonyl region of the uronic acids (González de Troconis et al., 2001), Figure 19.

Outstanding in this spectrum were the following resonances at:

- 179.47 ppm, assignable to residues of α-D-glucuronic acid, substituted by metals, previously described for gums of *A. niopoides* var colombiana and *P. saman* (León de Pinto et al., 1998b; 2002b).
- 174.07 ppm, due to acetyl groups, which corresponds to the intense signal that appeared at high field (20.44 ppm).

The spectral information obtained by 1D-$^{13}$C-NMR of the polysaccharide I, was further explored by two-dimensional spectroscopy of the original gum (González de Troconis et al., 2001). α

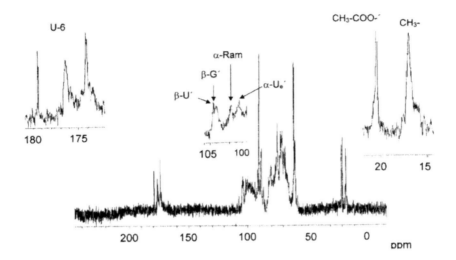

(Reprinted from *Ciencia*, 9, Nola González de Troconis, Maritza Martínez, Gladys León de y Alí Bahsas. Estudio químico y espectroscópico del polisacárido de la goma de *Cedrela odorata* [Chemical and spectroscopic study of *Cedrela odorata* gum polysaccharide], 285-293, Copyriht 2001, with permission from Ciencia).

Figure 19. $^{13}$C-NMR unidimensional del polisacárido I de la goma de *C. odorata*. U = ácidos urónicos G = galactosa,' = comprometido en el enlace glicosídico. [One-dimensional 13C-NMR of *C. odorata* polysaccharide I. U = uronic acids G = galactose, = compromised in the glycosidic bond].

The presence of 3-O-β-D-galactose was corroborated, by means of HMBC (H-1/C-3': 4.60/81.20 ppm). In the same way, two chemical environments were defined for 4-0-α-L-Ram (-CH$_3$/C-4': 1.3/82.38; 1.4/82.38 ppm). The presence of the α-D-glucuronic acid was also confirmed, through three scalar correlation, between H-3/C-1 (4.15/98.71 ppm), and the corresponding signals in HMQC H-3/C-3 (4.15/76.80 ppm) and H-1/C-1 (5.28/98.71 ppm).

## 4.2.2. Sterculia Gum Exudates

### 4.2.2.1. *Sterculia apetala* Gum

The gum of *Sterculia apetala* ("cacaito") can be included in this heteropolysaccharides B group, as a galacturonorhamnan. (Stephen et al., 1990). The presence of galactose was not evidenced. α-D-galacturonic acid was identified unlike was described for *C. odorata* gum. The presence of arabinose has not been reported for other gums of *Sterculia* (Verbeken et al., 2003; Brito et al., 2004) while that of xylose has been described for the *Sterculia striata* gum exudate (Brito et al., 2004).

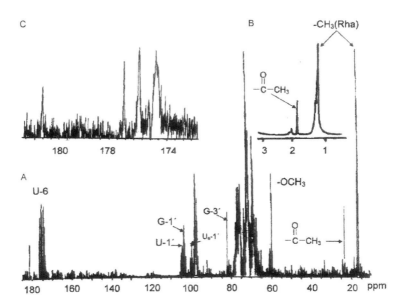

(Reprinted from *Food Hydrocolloids*, 20, Larrazábal Marvelys, Martínez Maritza, Sanabria Lilian, León de Pinto Gladys and Herrera Julio, Structural elucidation of the polysaccharide from *Sterculia apetala* gum by a combination of chemical methods and NMR spectroscopy, 908-913, Copyright 2006, with permission from Elsevier).

Figure 20. $^1$H and $^{13}$C-NMR spectrum of degraded gum A from S. apetala gum. A = 13CNMR spectrum (115-185 ppm), B = 1H-NMR spectrum and C = C-6 of uronic acid residues, G = β-D-galactose, U = uronic acids, Ue = 4-O-methyl-α-D-glucuronic acid.

The chemical studies revealed that a galucturonorhamnan (GalA/Rha: 1:3) is the core of *S. apetala* gum (Larrazabal et al., 2011). The constituent residues are linked together as follows: -->$^4$Rha$^1$-->$^4$Rha$^1$--> $^4$GalA$^1$--> $^4$Rha-->, forming a repetitive unit in the nucleus of the structure. The rhamnose is 4-O-linked unlike the *Sterculia urens* gum in which the rhamnose is 2-O-linked. (Verbeken et al., 2003). Chains of 2-O-L-rhamnose, substituted at O-3 or O-4, are present, while the D-xylopiranose and L-arabinofuranose are terminals in the structure (Larrazabal et al., 2011)

The 1D-Carbon-13 spectrum of the degraded A gum, Figure 20, showed intense signals, at high field, for the methyl of rhamnose (1.17; 1.2 ppm) and for acetyl groups (2.02; 2.08 ppm).

This spectrum showed, in the anomeric region, signals for:

- Reducing sugars: 97.17 ppm
- α-D-galacturonic acid: 98.04 ppm
- 4-O-methyl-α-D-glucuronic acid: 99.66; 100.43 ppm
- β-D-glucuronic acid: 103.65 ppm
- 3-O-β-D-galactose: 103.65 ppm

There were no evidences of 6-O-β-D-galactose as for other gums described above. At low field, stands out the signal, at 181.53 ppm, assignable to α-D-glucuronic acid substituted by metals (Larrazábal et al., 2001). Worthy of emphasis were the signals attributed to acetyl groups at high field.

The two-dimensional spectroscopy, HMQC and HMBC, of the polysaccharide I, corroborated the assignments performed for:

*Acetyl groups:*
H/OAc: 1.88/23.07 ppm            HMBC
1.88/174.71ppm                   HMBC
*4-O-methyl-α-D-glucuronic acid:*
H-1/H-2/H-3: 4.83/3.70/3.23 ppm  COSY
H-3/C-4: 3.23/82.30 ppm          HMBC
OMe/C-4: 3.33/82.30 ppm          HMBC

OCH₃/-OCH₃: 3.33/58.00 ppm        HMBC
α-L-rhamnopyranose
H/-CH3: 1.20/16.99 ppm            HMCC
CH3/C-5: 1.05/70.00               HMBC
H-5/C-5: 3.70/70.00               HMQC
H-1/C-5: 5.04/70.00               HMBC
H-1/C-1: 5.04/100.00              HMQC
*B-D-glucuronic acid:*
H-1/C-6: 4.55/75.00               HMBC
H-1/C-1: 4.55/103.10              HMQC

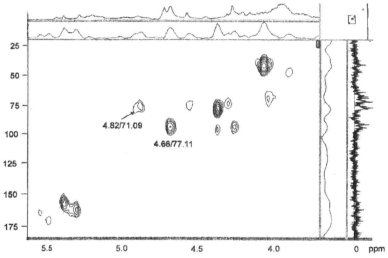

(Reprinted from *Food Hydrocolloids*, 20, Larrazábal Marvelys, Martínez Maritza, Sanabria Lilian, León de Pinto Gladys and Herrera Julio, Structural elucidation of the polysaccharide from *Sterculia apetala* gum by a combination of chemical methods and NMR spectroscopy, 908-913, Copyright 2006, with permission from Elsevier).

Figure 21. HSQC spectrum of degraded gum B from *S. apetala*.

The two-dimensional spectroscopic analysis of the degraded B gum, Figure. 21, on the other hand, confirmed the existence of the α-D-galacturonic acid substituted at O-4, according to the methylation analysis:

H-3/H-1: 4.28/3.65 ppm    COSY
H-3/C-3: 4.28/71.09 ppm   HMQC
H-3/C-6: 4.28/177 ppm     HMBC
H-4/C-4: 4.66/77.11 ppm   HMBC

### *4.2.3. Pereskia Gum Exudates*

#### *4.2.3.1. Pereskia guamacho* **Gum**

The gum of *Pereskia guamacho* (syn. *Rodocactus guamacho* (Web.) *Knuth*), Cactaceae, contains galactose (50%), arabinose (11%), xylose (2%), rhamnose (11%), and uronic acids (26%). It has an acidity content lower than that reported for the gum of *Sterculia apetala* but higher than the gums of *Acacia*. The uronic acids are represented by β-D-glucuronic acid, 4-O-methyl-α-D-glucuronic acid and α-D-galacturonic acid. The presence of this latter acid has been detected for gums of other Cactaceae (Nussinovitch, 2009) while that of xylose was identified for *A. tortuosa* (Martínez et al., 2015), *Spondias* (León de Pinto et al., 1996b), and *C. praecox* (León de Pinto et al., 1993; 1994a) gum exudates The highly positive specific rotation of the *P. guamacho* gum, is an indicator of the predominance of α-L-residues in the structure of the polymer. (León de Pinto et al., 1994b).

The backbone corresponds to a heteropolymer type B, constituted by galactose, rhamnose and uronic acids, analogous to that described for the gum of *Cedrela odorata*. The proportion of gal/rha is equal to 1.

The 1D-Carbon-13 spectroscopy of the degraded B gum allowed a better characterization of this heteropolymer, Figure 22 (Gutiérrez de Gotera et al., 2009). The spectrum showed four well-defined zones:

- *At high field:* There were identified signals for the–CH$_3$ groups of rhamnose, very intense in conformity with its high content, (16.61 ppm), -OCH$_3$ of 4-O-methyl-α-D-glucuronic acid, in two environments (56.15; 59.95 ppm), and free C-6 of β-D.galactose residues (60.78; 62.57.ppm).
- *In the glycosidic linkages region:* There were observed two signals, at 81.77: 82,00ppm, assigned to the C-3 linked of β-D-galactose

and/or the C-4 linked of α-L-rhamnose. These resonances were taken into consideration in the interpretation of the more complex spectra.

- *In the anomeric region:* Five intense signals were observed: C-1 of α-D-galacturonic acid (96.76; 97.97 ppm), C-1of β-L-rhamnose overlapped with C-1 of β-D-galactose (103.75 ppm), C-1 of 4-O-methyl-α-D-glucuronic acid (99.2 ppm) and C-1 of α-L-rhamnose (100.5 ppm).
- *At low field:* There were detected two signals which suggested two environments for the uronic acids residues: 176.68; 177.95 ppm.
- The spectrum of the original gum, Figure 23, is much more complex than the previous one, and did not allow to do unequivocal assignments in the region of the secondary carbons of the rings.

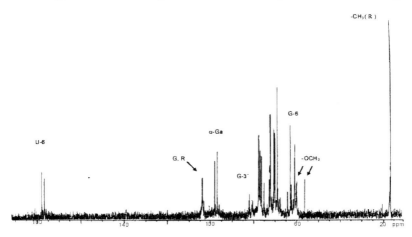

Legend: G-6=C-6 galactose G-6'=C-6 linked galactose G-3'=C-3 linked galactose Ag=galacturonic acid Ue=4-O-methyl derivative G=galactose R=rhamnose U=glucuronic acid Af=arabinofuranose

(Reprinted from *Ciencia*, 17, Omaira Gutiérrez de Gotera, Omaira Áñez de Servodio, Gladys León de Pinto, Ninoska Silva y J. Manuel Igartuburu, Rasgos estructurales relevantes del polisacárido presente en la goma de *Pereskia guamacho* [Relevant structural features of the polysaccharide present in the gum of *Pereskia guamacho*], 305 - 312, Copyright 2009, with permission from Ciencia).

Figure 22. Espectro de RMN-$^{13}$C de la goma degradada B de *Pereskia guamacho* [13 C-NMR spectrum of *Pereskia guamacho* degraded gum B].

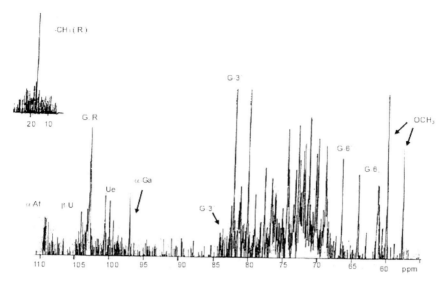

(Reprinted from *Ciencia*, 17, Omaira Gutiérrez de Gotera, Omaira Áñez de Servodio, Gladys León de Pinto, Ninoska Silva y J. Manuel Igartuburu Rasgos estructurales relevantes del polisacárido presente en la goma de *Pereskia guamacho* [Relevant structural features of the polysaccharide present in *Pereskia guamacho* gum] 305 - 312, Copyright 2009, with permission from Ciencia).

Figure 23. Espectro de RMN--$^{13}$C de la goma original de *Pereskia guamacho* [13C-NMR spectrum of *Pereskia guamacho* original gum].

Nevertheless, at high field, the signals for the methyl of rhamnose, -OCH$_3$ of the 4-O-methy-α-D-glucuronic acid and C-6 of 3-O-β-D-galactose were differentiated, as for the degraded B gum spectrum, Figure 22. There also exist spectroscopic evidences for 6-O-linked β-D-galactose(C-6 linked 66.51 ppm) and for C-5 of xylose (64.12 ppm). In the glycosidic linkage region, there were observed two intense signals, at 82.14, 82.47 ppm, which might correspond to the C-3 of 3-O-β-D-galactose and/or C-4 of 4-O-α-L-rhamnose.

In the anomeric zone, besides the signals described in the degraded B gum spectrum, there were also observed the resonances for α-L-arabinofuranose (109.2 ppm) and β-D-glucuronic acid (104.08ppm).

In the spectrum of the degraded gum A (not shown), there were differentiated the signals due to C-6 linked β-D-galactose (at 67.85 ppm), as evidence of a ramification point, and to C-5 of xylose at 65.17 ppm. In the anomeric region appeared the five types of linkages described for the original gum and, at a very low field, those relative to the C-6 of the uronic acids, highlighting the signal of higher intensity at 174.83 ppm. (suggesting the presence of acetyl groups) (León de Pinto et al., 1994b) as for *Sterculia* gums.

## 4.3. Type C Heteropolysaccharides

### 4.3.1. Cercidium praecox Gum

The polysaccharide gum of *Cercidium praecox* ("palo verde") belongs to the group C proposed by Stephen et al., 1990. The nucleus is a β (1-->4) xylan, unlike the structures of *Acacia* gums, which are galactans, but similar to the exudates of Araceae (Simas et al., 2004; 2006; 2014). Glucuronoxylans have also been described for the genus *Ananas* (pineapple) (Simas et al., 2006). The polysaccharide does not contain galactose or galacturonic acid as has been reported for other gums of *Cercidium* (Cerezo et al., 1969). The following neutral and acidic oligosaccharides, keys in the characterization of its structure, were isolated:

- α-D-glucuronic acid$^1$-------->$^4$xylose
- 4-O-methyl-α-D-glucuronic acid$^1$-------->$^2$xylose
- xylose$^1$-------->$^4$xylose
- xylose$^1$-------->$^4$xylose$^1$--------> $^4$xylose$^1$-------->$^4$xylose

In function of the chemical data, the spectrum of the original gum, Figure 24, showed, in the the glycosidic linkages region, the signals for C-4 of 4-O-β-D-xylose (75,35 ppm) and of 2,4-di-O-β-D- xylose (79.22 ppm) linked, through the C-2, to the 4-OMe-α-D-glucuronic acid according to chemical evidences (León de Pinto et al., 1994a). It has been reported, for the gum exudate of *Livistona chinensis* ("chinese palm"), a sequence which involves 4-O-methyl- α-D-GlcA (1---->2) β-D-Xylp(1--->4) β-D-Xylp (Maurer-Menestrina et al., 2003).

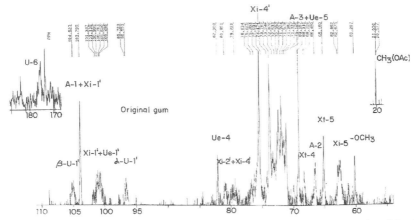

(Reprinted from *Carbohydrate Research*, 260, Gladys León de Pinto, Maritza Martínez and Carlos Rivas, Chemical and spectroscopic studies of *Cercidium praecox* gum exudate, 17-25, Copyright 1994, with permission from Elsevier).

Figure 24. $^{13}$C-NMR spectrum of *C. praecox* gum recorded with a Bruker AM-300, Xi internal β-D-xylose; Xt = terminal β-D-xylose; Ue = 4-O-methyl-α-D-glucuronic acid; U = glucuronic acid. 1,4-Dioxane (δ 66.67 ppm) was used as the internal standard.

On the other hand, there were identified, in the anomeric region, the signals for β-D-glucuronic acid (104.9 pppm), 4-O-methyl-α-D-glucuronic acid (100,50; 100,60 ppm), and α-D-glucuronic acid (96,55; 96,72 ppm). There were also distinguished the signals: at very low field (186.34 ppm), assigned to C-6 of α-D-glucuronic acid substituted by metals, as was described for gums of *A. niopoides* var. colombiana and *P. saman*, and at high field due to acetyl groups (20.58; 21.03 ppm). The existence of L-arabinopyranose chains Ara$_p$ 1----->3, (C-5: 66.67 ppm), was demonstrated, as well as for *A. tortuosa* gum, probably terminated in arabinofuranose and/or arabinopyranose residues (León de Pinto et al., 1994a).

The spectrum of the polysaccharide II was still complex. In general, all the signals present in the original gum spectrum, were observed. Nevertheless, there were not detected the signals, at 186.34 ppm, relative to C-6 of α-D-glucuronic acid substituted by metals, and of the acetyl groups (probably removed during the oxidation process).

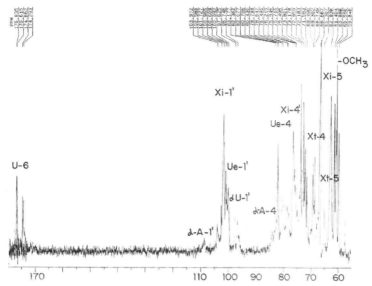

(Reprinted from *Carbohydrate Research*, 260, Gladys León de Pinto, Maritza Martínez and Carlos Rivas, Chemical and spectroscopic studies of *Cercidium praecox* gum exudate, 17-25, Copyright 1994, with permission from Elsevier).

Figure 25. $^{13}$C-NMR spectrum of polysaccharide II recorded with a Bruker AM-300. The same signals were observed in the spectrum of polysaccharide I..1,4-Dioxane ($\delta$ 66.67 ppm) was used as the internal standard.

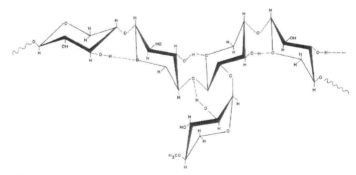

(Reprinted from *Carbohydrate Research*, 260, Gladys León de Pinto, Maritza Martínez and Carlos Rivas, Chemical and spectroscopic studies of *Cercidium praecox* gum exudate, 17-25, Copyright 1994, with permission from Elsevier).

Figure 26. Possible model for the core of Cercidium praecox gum. The (1-->4) xylan may be substituted either by $\alpha$-D-glucuronic acid or its 4-O-methyl ether.

In addition to this, the existence of resonances attributed to 3,4-di-O-substituted β-D- xylose residues, was confirmed, in accordance with the methylation analysis (C-3 80.07 ppm). Figure 25. A possible model was proposed, corresponding to a β-D-xylan (1---> 4), for a repetitive unit of the nucleus of the structure, based on the spectroscopic analysis and chemical evidences. Figure 26.

## Conclusion

The chemical and spectroscopic studies complement each other. In addition, the NMR spectroscopy allowed to deepen in the knowledge of the structures of these complex polymers

## Acknowledgments

The authors thank the editors of the journals belonging to the Elsevier group, Ciencia (Venezuela) and Revista Iberoamericana de Polímeros for their authorization to reuse the figures in this review.

## References

Anderson, D. M. W., Millar, J. R. A. & Weiping, W. (1991a). Gum arabic (*Acacia senegal*) from Niger. Comparison with other sources and potential agroforestry development. *Biochemical Systematics and Ecology*, 19, 447-452.

Anderson, D. M. W., Millar, J. R. A. & Weiping, W. (1991b). The gum exudates from *Combretum nigricans* gum, the major source of West African "gum combretum", *Food Additives and Contaminants*, 8, 423-436.

Aspinall, G. O. (1982). *The polysaccharides*. NY Academic Press.

Aspinall, G. O. & Bhatacharjee, A. K. (1970). Plant gums of the genus *Khaya*. Part IV: Major component of *Khaya ivorensis* gum. *Journal of the Chemical Society C: Organic*, *0,* 361-365.

Babu, A. M. & Shah, J. (1987). Unusual tissue complexes formed in association with traumatic gum cavities in the stems of *Bombax ceiba* L. *Annals of Botany*, *59,* 293-299.

Beltrán, O., León de Pinto, G., Martínez, M., Picton, L., Cozic, C., Le Cerf, D. & Muller, G. (2005). Fractionation and characterization of gum from *Acacia tortuosa*. Effect of enzymatic and alkaline treatments. *Carbohydrate Polymers*, *62,* 239-244.

Bock, I. C. & Pedersen, C. (1983). Carbon-13 NMR spectroscopy of monosaccharides. *Advances in Carbohydrates Chemistry and Biochemistry, 41,* 27-63.

Brito, A. C. F., Silva, D. A., Regina, C. M. P. & Feitosa, J. P. A. (2004). *Sterculia striata* exudate polysaccharide: characterization, rheological properties and comparison with *Sterculia urens* (karaya) polysaccharide. *Polymer International*, *53,* 1025-1032.

Carr, J. M. (1993). Hydrocolloids and Stabilizers. *Food Technology*, *97,* 100.

Cerezo, A. S., Stacey, M. & Webber, J. M. (1969). Some structural studies of brea gum (an exudate from *Cercidium australes* Jonhst.) *Carbohydrate Research*, *9,* 505-507.

Clamens, C., Rincón, F., Vera, A., Sanabria, L. & León de Pinto, G. (2000). Species widely disseminated in Venezuela which produces gum exudates. *Food Hydrocolloids*, *14,* 255-257.

Defaye, J. & Wong, E. (1986) Structural studies of gum arabic, the exudate polysaccharide from *Acacia senegal*. *Carbohydrate Research.*, *150,* 221-223.

Delgobo, C. I., Gorin, P. A. J., Tisher, C. A. & Iacomini, M. (1999). The free reducing oligossacharides of angico branco (*Anananthera colubrina*) gum exudates: an aid for structural assignments in the heteropolysaccharides. *Carbohydrate Research*, *320,* 167-175.

González de Troconis, N., Martínez, M., León de Pinto, G. & Bhasas, A. (2001). Estudio químico y espectroscópico del polisacárido de la goma de *Cedrela odorata*. [Chemical and spectroscopic study of the polysaccharide of Cedrela odorata gum]. *Ciencia.*, *9*, 285-293.

Gutiérrez de Gotera, O., Áñez de Servodio, O., León de Pinto, G. & Silva N y Igartuburu, J. M. (2009). Rasgos estructurales relevantes del polisacárido presente en la goma de *Pereskia guamacho* [Relevant structural features of the polysaccharide present in *Pereskia guamacho* gum] *Ciencia*, *17*(4), 305 – 312.

Joao, H. I., Jackson, G., Ravenscraft, M. & Stepehn, A. M. (1988). Structural aspects of 3-0-(α-D-galactopyranosil)-L-arabinose and the corresponding substituted L-arabinitol. *Carbohydrate Research*, *176*, 300-305.

Jones, J. K. N. & Smith, F. (1949). Gums and mucilages. *Advances in Carbohydrate Chemistry*, *4*, 243-283.

Larrazábal, M., Martínez, M., León de Pinto, G., El Kader, D. A., Herrera, J. y. & Bravo, A. (2011). Estudio de los principales enlaces glicosídicos presentes en la estructura de la goma de *Sterculia apetala* por degradación y análisis de metilación. [Study of the main glycosidic bonds present in the structure of *Sterculia apetala* gum by degradation and methylation analysis] *Ciencia.*, *19*, 136-141.

Larrazábal, M., Martínez, M., Sanabria, L., León de Pinto, G. & Herrera J. (2001). Structural elucidation of the polysaccharide from *Sterculia apetala* gum by a combination of chemical methods and NMR spectroscopy. *Food Hydrocolloids.*, *20*, 908-913.

Leal, M., Martínez, M., Beltrán, O., León de Pinto, G., Picton, L., Didier, L. C. & Oulyadi, H y Guilhaudis L. (2010). Estudio de la Interacción carbohidrato-proteína en la goma de *Spondias mombin* por cromatografía de exclusión (SEC) y RMN. [Study of carbohydrate-protein interaction in *Spondias mombin* gum by exclusion chromatography (SEC) and NMR] *Revista Iberoamericana de Polímeros.*, *11*, 485-504.

León de Pinto, G. & González de Troconis, N. (1989). Espectros de RMN de la goma de *Albizia lebbeck* y de sus productos. [NMR spectra of Albizia lebbeck gum and its products.] *Acta Científica Venezolana.*, *40*, 335-340.

León de Pinto, G., Martínez, M., Beltrán, O., Rincón, F., Igartuburu, J. M. & Rodríguez, F. (2000). Structural investigation of the polysaccharide of *Spondias mombin*. *Carbohydrate Polymers.*, *43*, 105-112.

León de Pinto, G., Martínez, M., Mendoza, J. A., Ávila, D., Ocando, E. & Rivas, C. (1996a). Structural study of the polysaccharide isolated from *Spondias purpurea* gum exudate. *Carbohydrate Researc*, *290*, 97-103.

León de Pinto, G., Martínez, M., Ocando, E. & Rivas, C. (2001b). Relevant structural features of the polysaccharide from *Pithecellobium mangense* gum exudate. *Carbohydrate Polymers.*, *46*, 261-266.

León de Pinto, G. & González de Troconis, N. (1989). Espectros de R.M.N. de la goma de *Albizia lebbeck* y de sus productos degradados. Aplicación a su elucidación estructural. [NMR spectra. of *Albizia lebbeck* gum and its degraded products. Application to its structural elucidation] *Acta Científica Venezolana*, *40*, 335-340.

León de Pinto, G., González de Troconis, N., Martínez, M., Vera, A., Rivas, C. & Ocando, E. (1996b). Composition of three Meliaceae gum exudates. *Ciencia*, *4*, 45-52.

León de Pinto, G., Martínez, M., de Bolaño, L. M. & Rivas, C. (1998a). The polysaccharide gum from *Acacia tortuosa*. *Phytochemistry*, *47*, 53-56.

León de Pinto, G., Martínez, M. & Rivas, C. (1994a). Chemical and spectroscopic studies of *Cercidium praecox* gum exudate. *Carbohydrate Research.*, *260*, 17-25.

León de Pinto, G., Martínez, M. & Sanabria, L. (2001a). Structural features of the polysaccharide gum from *Acacia glomerosa*. *Food Hydrocolloids.*, *15*, 461-467.

León de Pinto, G., Martínez, M., Beltrán, O., Rincón, F., Clamens, C., Igartuburu, J. M., Guerrero, R. & Vera, A. (2002). Characterization of polysaccharides isolated from gum of two Venezuelan specimens of *Albizia niopoides* var. colombiana. *Ciencia.*, *10*, 382-38.

León de Pinto, G., Martínez, M., Gutiérrez de Gotera, O. & Rivas C y Ocando, E. (1998b). Structural study of the polysaccharide isolated from *Samanea saman* gum. *Ciencia*, *6*(3), 191-199.

León de Pinto, G., Martínez, M., Gutiérrez de Gotera, O., Vera, A., Rivas, C. & Ocando, E. (1995). Comparison of two *Pithecellobium* gum exudates. *Biochemical Systematics and Ecology*, 23, 849-853.

León de Pinto, G., Paz de Moncada, N., Martínez, M., Gutiérrez de Gotera, O., Rivas, C. & Ocando, E. (1994b). Composition of *Pereskia guamacho* gum exudates. *Biochemical Systematics and Ecology*, 22, 291-295.

León de Pinto, G., Rodríguez, O., Martínez, M. & Rivas, C. (1993). Composition of *Cercidium praecox* gum exudates. *Biochemical Systematics and Ecology.*, 21, 297-300.

León de Pinto, G., Sanabria, L., Martínez, M., Beltrán, O. & Igartuburu, J. M. (2002a). Structural elucidation of proteic fraction isolated from *Acacia glomerosa* gum. *Food Hydrocolloids*, 16, 599-603.

López-Franco, Y., Goycolea, F. & Váldez N y Calderón de la Barca, A. M. (2006). Goma mezquita una alternativa de uso industrial [Mezquite gum an alternative for industry use], *Interciencia*, 31 (3), 183-189.

Martínez, M., Beltrán, O., Rincón, F. & León de Pinto G e Igartuburu, J. M. (2015). New structural features of *Acacia tortuosa* gum exudate. *Food Chemistry*, 182, 105-110.

Martínez, M., León de Pinto, G., Bozo de González, M., Herrera, J. & Guilhaudis, L. (2008). New structural features of *Spondias purpurea* gum exudate. *Food Hydrocolloids*, 22, 1310-1314.

Martínez, M., León de Pinto, G., Rivas, C. & Ocando, E. (1996a). Chemical and spectroscopic studies of the gum polysaccharides from *Acacia macracantha*. *Carbohydrate Polymers*, 29, 247-252.

Martínez, M., León de Pinto, G., Alvárez, S., González de Troconis, N., Ocando, E. & Rivas, C. (1996b). Composition and properties of *Albizia lebbeck* gum exudate. *Biochemical Systematics and Ecology*, 23, 843-848.

Martínez, M., León de Pinto, G., Sanabria, L., Beltrán, O., Igartuburu, J. M. & Bhasas, A. (2003). Structural features of an arabinogalactán gum exudates from *Spondias dulcis* (Anacardiaceae). *Carbohydrate Research*, 338, 619-624.

Maurer-Menestrina, J., Sassaki, G. L., Simas, F. F., Gorin, P. A. & Iacomini, M. (2003). Structures of a highly substituted β-xylan of the gum exudate of the palm *Livistona chinensis* (Chinese fan). *Carbohydrate Research*, *1338*, 1843-1850.

Mc Intyre, D. D., Cer, H. & Vogel, H. J. (1996). Nuclear magnetic resonance studies of the heropoysaccharides alginates, gum Arabic and gum xanthan. *Starch*, *48*, 285-290.

Nussinovitch, A. (2010). Plant gum exudates of the world. *Sources, Distribution, Properties and Applications.* CRS Press. FL, USA.

Randall, R. C., Phillips, G. O. & Williams, P. A. (1988). Fractionation and characterization of gum from *Acacia senegal. Food Hydrocolloids*, *3*, 65-75.

Rincón, F., Muñoz, J., Ramírez, P., Galán, H. & Alfaro, M. C. (2014). Physicochemical and rheological characterization of *Prosopis juliflora* seed gum aqueous dispersions. *Food Hydrocolloids*, *35*, 348-357.

Simas, F. F., Gorin, P. A. J., Guerrini, M., Naggi, A., Sassaki, G. L. & Delgobo, C. L. (2004). Structure of a heteroxylan of gum exudate of the palm *Scheelea phalerata* (uricuri). *Phytochemistry*, *65*, 2347–2355.

Simas, F. F., Maurer-Menestrina, J., Reis, R. A., Sassaki, G. L., Iacomini, M. & Gorin, P. A. J. (2006). Structure of the fucose-containing acidic heteroxylan from the gum exudate of *Syagrus romanzoffiana* (Queen palm). *Carbohydrate Polymers*, *63*, 30–39.

Simas, F. F., Barraza, R., Maria-Fereira, D., Smiderle, F., Carbonero, E., Sassaki, G. L., Iacomini, M. & Gorin, P. A. (2014). Glucurono-arabinoxylan from coconut palm exudate: Chemical structure and gastroprotective effect. *Carbohydrate Polymers*, *107*, 65-71.

Sims, I. M. & Furneaux, R. H. (2003). Structure of the exudate gum from *Meryta sincairii. Carbohydrate Polymers*, *52* (6), 423-431.

Singh, B. R., Dubey, S. & Siddiqui, M. Z. (2015). *Antimicrobial activity of natural edible gums World Journal of Pharmaceutical Sciences ISSN (Print)*, 2321-3310, ISSN (Online), 2321-3086.

Stephen, A. M. (1983). Structure and properties of exudate gums. In: *The polysaccharides* (GO Aspinall) NY Academic Press.

Stephen, A. M., Churms, S. C. & Vogt, D. C. (1990). Exudate gums. *Methods in Plant Biochemistry*, *2*, 483-522.

Tisher, C. A., Gorin, P. A. J. & Iacomini, M. (2002). The free reducing oligosaccharides of gum Arabic: aids for structural assignments in the polysaccharides. *Carbohydrate Polymers*, *47*, 151-158.

Utille, J. P., Kovac, P., Sauriol, F. & Perlin, A. S. (1986). NMR spectra of aldobiuronic and aldobiouronic acid derivatives related to 4-O-methyl-D-glucuronoxylans. *Carbohydrate Research*, *154*, 251-258.

Verbeken, D., Dierreckx, S. & Dewettinck, K. (2003). Exudate gums: occurrence, production, and applications *Applied Microbiology and Biotechnology*, *63*, 10-21.

Yadav, M. P. & Nothangel, E. N. (2006). Chemical and composition of an effective emulsifier subfraction of gum Arabic. Chapter 16., pp. 243-254, In: *Advances in Biopolymers*. ACS Books. Washington DC. USA.

In: *Acacia*
Editor: Aide Matheson

ISBN: 978-1-53614-237-2
© 2018 Nova Science Publishers, Inc.

*Chapter 5*

# RECONSTRUCTION OF ANCESTRAL CHARACTER STATES IN NEOTROPICAL ANT-*ACACIAS*

*Jessica Admin Córdoba de León*[*]
*and Sandra Luz Gómez Acevedo*[±]*, PhD*

Unidad de Morfología y Función, Facultad de Estudios Superiores Iztacala, Universidad Nacional Autónoma de México, Tlalnepantla, Estado de México, México

## ABSTRACT

The footprints left by evolution in the distribution of characters among current organisms have been one of the main tools in the study of organic evolution. The reconstruction of ancestral character states offers the possibility of knowing the changes suffered by characters in a species throughout evolutionary time; of understanding the origin of the adaptations to which they give rise, and comprehending their function as well as recognizing the way in which the symbiotic interactions that they

---

[*] Corresponding Author Email: mar.9410@gmail.com;
[±] Corresponding Author Email: sanluza@unam.mx

sustain with other species influence such attributes. The relationships between plants and ants provide numerous examples of mutualism, one of them is the obligate and highly specialized interaction between ants and *Acacias*. Neotropical ant-*Acacias* are characterized by the presence of extrafloral nectaries, domatia and beltian bodies; traits directly related to their mutualistic association with ants. These traits when analyzed under the perspective of the reconstruction of ancestral character states show that during the transition from the external group to the myrmecophile group the amount of extrafloral nectaries per petiole increased and their shape changed from circular to columnar and canoe-shaped. The stipules grew in length resulting in domatia and cyanogenic glycosides against herbivory were lost. Furthermore, in relation to other morphological characteristics, the leaves were lengthened in a first step and then became smaller. The number of pinnae pairs and pinnules per leaf decreased, while both the length and width of the pinnules increased, as well as the petiole maximum length. Thus, the selection pressure that symbiotic relationships exert on interacting species is reflected in a systematic set of their attributes, which, in our case, were developed in an environmental scenario of warm-humid conditions prevalent in the Neotropics during the Late Miocene and Early Pliocene.

## 1. INTRODUCTION

Evolution leaves its footprints in the distribution of characters among current organisms, footprints that have been one of the main tools for the study of organic evolution. This historical and comparative approach is especially important for the study of adaptive characters, since only in a of phylogenetic reference frame; fundamental concepts such as convergence, parallelism and adaptive radiation become significant (Pagel et al. 2004; Ronquist 2004; Schultz et al. 1996).

Yet, to understand evolution, it is necessary to know the characteristics of current organisms. As well as those characteristics present in their ancestors. A popular alternative is to reconstruct the evolutionary history of an individual's characters. Or to infer the characteristics of the ancestor, based on a phylogenetic tree (Cunningham et al. 1998; Mooers et al. 1999; Pagel et al. 2004; Ronquist 2004).

## 1.1. Reconstruction of Ancestral Character States

The estimation of the characteristics of ancestral hypothetical species in the phylogenetic context is a domain of comparative phylogenetic biology. Evolutionary models and the set of assumptions provided by the comparative method can be used to discover, in a way that could be described as "statistical paleontology", the diversity of biological characters (nucleotide or amino acid sequences; ecological, phenotypic or biogeographic traits) manifested in ancestral species, as well as the nature of the underlying evolutionary processes (Harvey & Pagel 1991; Omland 1999; Pagel 1999a, b; Revell 2014; Royer-Carenzi et al. 2013; Vanderpoorten & Goffinet 2006).

This is an attractive idea, because it offers us the possibility to glimpse the past, to discover how characters evolve and to understand their function. To date, different methods have been developed to address the problem of ancestral state reconstruction. Two of the most discussed and used methods are maximum parsimony and maximum likelihood. The method of maximum parsimony is very effective when the degree of divergence between species is low and is particularly successful in the analysis of discrete characters that vary dichotomically (Harvey & Pagel 1991). The character states of ancestral taxa are inferred independently for each topology, under the assumption that changes between each state occur in all possible directions. The total number of character state changes that explain the complete evolutionary process for a given topology is then registered. This computation is done for all potentially correct topologies and the one that requires the least number of changes; that is, the lowest number of independent sources of the shared characters; is chosen as the best tree. In this way, parsimony is built around the proposition that the "best tree" is the one that describes the evolution of any particular set of characters using as few evolutionary changes as possible (Harvey & Pagel 1991; Nei & Kumar 2000; Wiley & Lieberman 2011).

In the maximum likelihood method, the tree that has the highest probability of producing the observed data set; given a specific evolutionary model adopted by the researcher, as well as the topology of the tree and the length of branches between nodes; is chosen. The estimators used have the property of being consistent in a wide range of evolutionary rates and phylogenetic topologies, ensuring that as more data is collected, the results will be more accurate and will present smaller variances (Harvey & Pagel 1991; Nei & Kumar 2000). In this way, a point estimate of the best tree is produced given the criterion of the adjusted model. That is, the adjusted data in a defined topology and its parameters are the "best model" (Wiley & Lieberman 2011).

Despite its great importance, the reconstruction of ancestral character states poses a tremendous challenge due to the inherent limitations of data (Omland 1999). Most of these limitations are related to the fact that existing taxa represent only a subset of the total evolutionary diversity of any clade, and, therefore, any conclusions based solely on existing taxa should be taken with caution. The inferences made through the reconstruction of ancestral character states necessarily make an implicit but problematic assumption that the distribution of the characters among the living taxa faithfully records the evolution patterns of the characters throughout the history of that group (Finarelli & Flynn 2006).

However, the wide range of questions that can be addressed through the inference of ancestral states or change trajectories of key characters in phylogenetic trees is fascinating. The reconstruction of the probable ancestral states of organisms has been used to reveal homologies between characters; study morphological conservatism and homoplasia; check if there is neutral evolution or functional divergence in any specific system; detect correlated transitions between characters in coevolutionary lineages; examine the origin of adaptations; evaluate variations in diversification rates; re-evaluate past classifications; investigate the ancient characteristics of life on Earth, as well as to evaluate ecological and evolutionary hypotheses (Collins et al. 1994; Cunha et al. 2015; Cunningham et al. 1998; de Oliveira et al. 2014; Givnish et al. 2005, Horn et al. 2009; Nurk et al. 2013; Pagel et al. 2004; Ricklefs 2007; Ronquist 2004; Royer-Carenzi et al. 2013; Schäffer et al. 2010; Simpson 2010; Soltis et al. 2013; Webster et al. 2012; Wu et al. 2015).

In this way, through tools such as the reconstruction of character states of the most recent common ancestor of a specific group it is possible to respond to a number of questions of an evolutionary nature, among them, the influence of symbiotic relationships, such as myrmecophily, on the evolution of the attributes of interacting species, a topic that we will develop here in detail.

## 1.2. Plant-Ant Mutualism

All organisms develop in a complex network of ecological interactions that significantly influence the evolution of their characters and, therefore, the diversification and speciation of their lineages. Mutualism is an association between organisms of different species that increases the fitness of each interacting individual (Bronstein 1994; Chenuil & Mckey 1996; Fiala et al. 1999; Howe & Westley 1989).

Relations between plants and ants provide numerous examples of mutualism, ranging from opportunistic, facultative and generalist associations, to highly specialized and obligate interactions such as the symbiosis between ants and myrmecophytes (Bronstein et al. 2006; Heil 2008; Heil & McKey 2003; Mayer et al. 2014; Rico-Gray & Oliveira 2007). The latter possess a great variety of adaptive ecological and morphological features, most of which seem to be related to the mutual association with ants. Among the most representative examples of these plants are myrmecophyte *Acacias*.

## 1.3. Myrmecophily in *Acacias*

The genus *Acacia* Miller *sensu lato* groups approximately 1600 species, distributed in three subgenera: subg. *Acacia* (= *Vachellia*), subg. *Aculeiferum*, and subg. *Phyllodineae* (= *Acacia*) (Lewis et al. 2015; Miller & Seigler, 2012). Of all these species, only four African and 15 Neotropical; all belonging to the *Acacia* subgenus; are myrmecophytes (Janzen 1974; Lewis et al. 2005; Rico-Arce 2003) and represent the only known taxon whose myrmecophyte species appear both in the Neotropical and Paleotropical regions (Davidson & McKey 1993).

The 15 species of Neotropical myrmecophytes (*A. allenii, A. cedilloi, A. chiapensis, A. cookii, A. cornigera, A. collinsii, A. gentlei, A. globulifera, A. hindsii, A. hirtipes, A. janzenii, A. mayana, A. melanoceras, A. ruddiae* and *A. sphaerocephala*) are shrubs or trees with bipinnate composite leaves that are characterized by having: (1) large, hollow spinescent stipules (domatia) that the ants use as shelter; (2) leaflet tips modified into yellow corpuscles composed of lipids (1-10% of dry weight); proteins and free amino acids (8-14% of dry weight); carbohydrates (3-11% dry weight) and water (18-24% dry weight) called Beltian bodies, which are harvested by ants and intended primarily for feeding the larvae; (3) continuous production of leaves, guaranteeing a constant source of food for the ants, since Beltian bodies, once cut, do not regenerate; (4) enormously enlarged petiolar nectaries that continuously secrete nectar, which is the main source of water and carbohydrates for adult ants and (5) loss of most of the chemical characters (alkaloids and cyanogenic glycosides) that protect plants from most herbivores (Heil et al. 2004; Janzen 1974; Rehr et al. 1973).

Molecular studies support the monophyly of the subgenus *Acacia*, which is subdivided into several informal groups that also resolve as monophyletic. Among the informal groups are the Neotropical myrmecophile *Acacias* (the object of study in this chapter), with an estimated divergence time of 5.44 ± 1.93 million years, which are intimately related to the *Acacia macracantha* group (4.57 ± 2.14 million years) by common ancestry, whose probable ancestor may have existed during the Miocene, approximately 6.92 ± 2.49 million years ago (Gómez-Acevedo et al. 2010).

## 1.4. Justification and Objectives

Although previously good diagnoses and descriptions of the Neotropical *Acacia-Pseudomyrmex* mutualist system have been made (Janzen 1966, 1974), no previous study has managed to clarify how the ancestor of Neotropical ant-*Acacias* was like, or to elucidate the history of change of the characters; as well as when and under which environmental and historical conditions the characters of interest may have arisen and evolved. In this sense, the reconstruction of ancestral states of the myrmecophile group of Neotropical *Acacias* can be one of the possible approximations in the search for an answer to this question, and eventually in the understanding of this phenomenon.

Due to the aforementioned, the present study has as main objective to know the possible evolutionary changes in the states of the most representative vegetative characters in the myrmecophile group of Neotropical *Acacias*, considering the *A. macracantha* group as an external group. It is also proposed, by comparing the ancestral and derived states, how these characters have evolved under the influence of the mutualist association with ants, including the paleoenvironmental context.

## 2. MATERIALS AND METHODS

A bibliographic search was carried out including information related to vegetative characters for both the species of the myrmecophile group and the *Acacia macracantha* group. The characters as well as their coding are shown in Table 1. For the reconstruction of the character states, the ultrametric phylogenetic tree proposed by Gómez-Acevedo et al. (2010) was used. The nomenclature used here is as follows: node A) includes the common ancestor of the myrmecophile group and the external group; node B) includes the myrmecophile group; and node C) includes the *A. macracantha* group (Figure 1). For the analysis, the characters were discriminated depending on their discrete or continuous nature and the continuous variables were analyzed independently from the discrete variables. In the case of discrete characters, 2 to 3 states were determined for each character depending on their attributes.

The resulting matrices were used to reconstruct the ancestral character states. Considering that different analytical methods have different advantages and limitations, in this study both the maximum parsimony method and the maximum likelihood method were used for the reconstruction of character states. Both analyzes were performed in the

Mesquite 3.04 program (Maddison & Maddison 2015). In the reconstruction carried out with the maximum parsimony method, the character states were treated as disordered, that is, Fitch's parsimony was used. In addition, the 1-parameter Markov k-state model (Mk1) was used for the reconstruction under the maximum likelihood method, with the same probability for any change of character. With the obtained results, the plesiomorphic and apomorphic character states for the myrmecophile group were established and compared with each other in order to appreciate the changes that these characters suffered during the transition from the external group to the myrmecophile group.

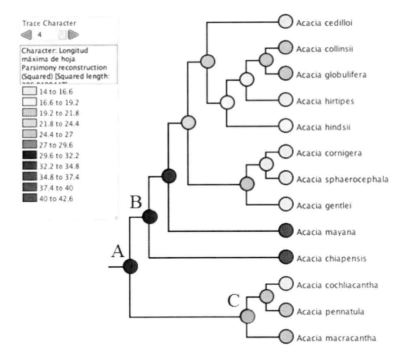

Figure 1. Maximum leaf length reconstruction of myrmecophilous *Acacias* and external *A. macracantha* group by maximum parsimony analysis of continuous characters. Node A includes the common ancestor of the ant-*Acacias* group and the external group.

At the same time, a literature search and synthesis was carried out on the paleoenvironmental context in which the species of both the myrmecophile and the *A. macracantha* groups were probably developed. For this, all possible reports that corresponded to the characterization or qualitative description of what happened in the Neotropic during the time period between the end of the Miocene and the beginning of the Pliocene, were traced.

**Table 1. Characters and their coding in this study**

| Characters | Character states |
|---|---|
| 1. Habit | 0: tree, 1: tree-bush * |
| 2. Maximum height (m) | 0: 1 to 6; 1: 6.1 to 12; 2: more than 12 |
| 3. Maximum leaf length (cm) | 0: 14 to 20; 1: 20.1 to 30; 2: more than 30 |
| 4. Pinnae pairs per leaf (minimum) | 0: 1 to 5; 1: 5.1 to 10; 2: more than 10 |
| 5. Pinnae pairs per leaf (maximum) | 0: 10 to 20; 1: 20.1 to 30; 2: more than 30 |
| 6. Pinnule pairs per pinna (minimum) | 0: 5 to 15; 1: 15.1 to 25; 2: more than 25 |
| 7. Pinnule pairs per pinna (maximum) | 0: 20 to 30; 1: 30.1 to 50; 2: more than 50 |
| 8. Minimum pinnule length (mm) | 0: 0.5 to 5; 1: 5.1 to 10; 2: more than 10 |
| 9. Maximum pinnule length (mm) | 0: 0.5 to 10; 1: 10.1 to 20; 2: more than 20 |
| 10. Minimum pinnule width (mm) | 0: 0.1 to 1; 1: more than 1 |
| 11. Maximum pinnule width (mm) | 0: 0.1 to 1.1; 1: 1.2 to 2.9; 2: more than 2.9 |
| 12. Maximum petiole length (mm) | 0: 10 to 15; 1: 15.1 to 20; 2: more than 20 |
| 13. Form of extrafloral nectaries | 0: volcano; 1: canoe; 2: columnar |
| 14. Number of extrafloral nectaries per petiole | 0: 1; 1: 2 to 7; 2: 4 to 14 |
| 15. Spinescent stipules | 0: not modified; 1: modified |
| 16. Length of spinescent stipules (mm) | 0: less than 50; 1: 50 to 75; 2: more than 75 |
| 17. Secondary metabolites | 0: without cyanogenics; 1: S-proacacipetalin; 2: R-epiproacacipetalin |

* = indicates that the species that have a life-form corresponding to the tree-shrub character state can develop by presenting any of the two habits; m = meters; cm = centimeters

## 3. RESULTS AND DISCUSSION

### 3.1. Analysis of Used Methods

The reconstruction of the character states of the discrete variables in the most recent phylogenetic tree (Gómez-Acevedo et al. 2010) did not result in any conflict between the maximum parsimony and maximum likelihood analyzes, that is, the character states of the resulting topologies in both analyzes coincide with each other. However, some character states of the maximum parsimony analysis were ambiguous compared to those obtained through the maximum likelihood analysis, this is the case of the present secondary metabolites.

Given that the results of both analyzes yielded the same information and that the maximum likelihood method, unlike that of maximum parsimony, provides the probabilistic values associated with each possible character state, from now on, only the results obtained with maximum likelihood will be referred to. Also, for continuous variables, only the maximum parsimony analysis could be executed in the program.

Reconstructions based on maximum likelihood analysis use an explicit character evolution model to estimate the probabilities of all possible ancestral states in each node of the tree (Cunningham et al. 1998). One of the strengths of this method is that it reveals the amount of uncertainty in the reconstruction of each character state, even when these states are unequivocally reconstructed under the maximum parsimony method (Nepokroeff et al. 2003). In this study, the reconstruction performed through the maximum likelihood analysis was consistent with that obtained by means of maximum parsimony. However, some of the results achieved through maximum parsimony were ambiguous compared to those obtained through maximum likelihood, which confirms some of the advantages of the latter method.

It should be noted that in reconstructions it is necessary to consider the directionality of the character-state change trajectories, since the evolutionary pattern influenced by positive selection is often unique and may be different from the one in which there is no adaptive change. This tendency may affect the maximum likelihood analysis, so that the posterior probabilities calculated by this approximation may not be realistic (Harvey & Pagel 1991; Nei & Kumar 2000). However, in the analysis conducted in this study, the probability of making an error due to this type of trends was reduced when we simultaneously considered the results of the parsimony analysis.

In the analysis carried out through the maximum parsimony method, the Fitch model was used; a form of analysis, which, as mentioned before, considers character states as disordered. It is important to highlight that although for all the characters analyzed here, the evolutionary model to which each of them adheres is unknown, the use of Fitch's parsimony is justified under the assumption that (1) the purpose of the present study was simply to explore the possible results of such reconstruction and that (2) Fitch's parsimony is effective in the treatment of multi-state morphological characters (Harvey & Pagel 1991; Wiley & Lieberman 2011).

Thus, both the maximum parsimony method and the maximum likelihood method have certain advantages and disadvantages, which is why, according to our observations, the best approach for ancestral state reconstruction consists of considering both analyzes simultaneously.

## 3.2. Reconstruction of the Characteristics of the Common Ancestor of the Ant-*Acacias* and External Groups

The probable ancestor had a habit corresponding to the tree-shrub combination and had a maximum height of 11.8 to 14.4 m. However, it was difficult to identify the plesiomorphic and the apomorphic character, since both the lowest and the highest values of the maximum height range that these species can reach are represented in the external group. However, the most frequent size range is that of 9.2 to 11.8 m, therefore, it is possible that this measure corresponds to the plesiomorphic character.

According to Harvey & Pagel (1991), the most common character state in a group can also correspond to the derived state. For this reason, a second option to consider is to choose the state that the external group presents as ancestral, a value that in this analysis corresponds to the interval between 11.8 and 14.4 m. In this way, the derived states may correspond either to the 4-9.2 m range, or to the 4-11.8 m range (Table 2).

**Table 2. Observed character states in the common ancestor of the myrmecophile group and the *A. macracantha* group (node A), in the ancestor of the myrmecophile group (node B) and in the ancestor of the *A. macracantha* group (node C)**

| Character | Node A | Node B | Node C |
|---|---|---|---|
| Habit | tree-bush (0.6143) | tree-bush (0.9248) | tree (0.6803) |
| Maximum height (m) | 11.8 to 14.4 | 9.2 to 11.8 | 11.8 to 14.4 |
| Maximum leaf length (cm) | 29.6 to 32.2 | 32.2 to 34.8 | 24.4 to 27 |
| Pinnae pairs per leaf (minimum) | 5 to 10 (0.8928) | 5 to 10 (0.9547) | 5 to 10 (0.9220) |
| Pinnae pairs per leaf (maximum) | more than 30 (0.7483) | more than 30 (0.7087) | more than 30 (0.8665) |
| Pinnule pairs per pinna (minimum) | 15 to 25 (0.6942) | 15 to 25 (0.8292) | 15 to 25 (0.6786) |
| Pinnule pairs per pinna (maximum) | more than 50 (0.5305) | more than 50 (0.5521) | more than 50 (0.5436) |
| Minimum pinnule length (mm) | 2.64 to 3.56 | 3.56 to 4.48 | 1.72 to 2.64 |
| Maximum pinnule length (mm) | 6.52 to 8.58 | 8.58 to 10.64 | 4.46 to 6.52 |
| Minimum pinnule width (mm) | 0.6 to 0.75 | 0.9 to 1.05 | 0.45 to 0.6 |
| Maximum pinnule width (mm) | 1.62 to 1.96 | 1.96 to 2.3 | 1.28 to 1.62 |
| Maximum petiole length (mm) | 15 to 17.5 | 17.5 to 20 | 12.5 to 15 |
| Form of extrafloral nectaries | circular (0.5318) | columnar or canoe (0.4731, 0.3357) | circular (0.9280) |
| Number of nectaries per petiole | 1 to rarely 2 (0.9274) | 1 to rarely 2 (0.8722) | 1 to rarely 2 (0.9956) |
| Spinescent stipules | ? | modified (0.9770) | unmodified (0.9771) |
| Spinescent stipule length (mm) | 46 to 55 | 64 to 73 | 28 to 37 |
| Secondary metabolites | S-proacacipetalin (0.7340) | S-proacacipetalin (0.6951) | S-proacacipetalin (0.8480) |

The decimals indicated in parentheses correspond to the probabilities of finding a determined character state in the corresponding nodes

Following this line of analysis, in the transition from the *A. macracantha group* to the myrmecophile group the species tended to diminish in the value of that character. That is, the upper limit of the maximum height range went from 9.2-14.4 m to 4-11.8 m. This result contrasts with the description of Janzen (1966), who reports that Neotropical ant-*Acacias* vary from small shrubs to trees 30 m high. However, according to the records compiled here, the highest species of the myrmecophile group is *A. gentlei*, reaching 22.2 to 24.8 m, whereas the rest of the species in this group reach a maximum height of 9.2 to 11.8 m. It is necessary to point out that it is unknown if there is any direct relationship between the reduction in maximum height that these species can reach and myrmecophily.

Regarding leaf maximum length, the analysis showed values from 29.6 to more than 32.2 cm. Also, the probable ancestor presented from 5 to more than 30 pairs of pinnae per leaf and from 15 to more than 50 pairs of pinnules per pinna. Their pinnules measured from 2.64 to 8.58 mm in length and from 0.6 to 1.96 mm in width. Its petioles had a maximum length of 15 to 17.5 mm. In other words, according to the results of the present study, during the transition from the myrmecophile group to the external group, the leaves enlarged in a first step (40 to 42.6 cm) and then became even smaller (14 to 21.8 cm) compared to its initial size (24.4 to 27 cm, Table 2). This is contradictory with what was previously reported, because according to Janzen (1966, 1974), the ant-*Acacias* have the longest leaves of this genus, since they extend far beyond the tips of the domatia and can be easily removed from the tree without touching said structures.

According to this same author, large leaves are useful in cloudy climates (which occur more commonly in wet than in dry locations), in addition to being favored by the absence of leaf water loss in this type of habitats. In contrast, in drier habitats, many of the larger leaves fall during the dry season, and then axillary groups of very small leaves provide the colony with nectar and Beltian bodies, essential products that prevent the colony from dying or migrating to another individual. Very few myrmecophile plants with large leaves are produced during the dry season. Thus, the species that inhabit the driest places (*A. collinsii*, *A. cornigera* and *A. hindsii*) have the smallest leaves.

In the same way, both the number of pinnae pairs per leaf and the pinnule number per pinna decreased during this transition. That is, from a range of 5 to more than 30 pairs of pinnae per leaf and from 15 to more than 50 pairs of pinnules per pinna (300 to more than 6000 pinnules per leaf) they went on to possess from 1 to a maximum of 30 pairs of pinnae per leaf and from 5 to a maximum of 50 pairs of pinnules per pinna (10 to maximum of 6000 pinules per leaf). On the contrary, an increase in pinnae length and width is observed during the transition from the external group to the myrmecophile group. The pinnules went from an area of 0.8 to 6.52 mm in length and 0.3 to 1.28 mm in width, to a range of 2.64 to 16.82 mm in length and 0.6 to 3.66 mm in width (Table 2).

In other words, the ant-*Acacias* have smaller leaves and a smaller amount of pinnae per leaf and of pinnules per pinna than the species of the *A. macracantha* group, but also have longer and wider pinnules in comparison with these species. This pattern of change is unprecedented in the literature and it is possible that it is not only related to the abiotic context, but also to the biological environment of Neotropical ant-*Acacias*.

Regarding pinnule length and width, Janzen (1974) proposed that the leaves of myrmecophile *Acacias* are shaped, in general, by larger pinnules than those present in the rest of the Neotropical species, but he did not delve into the possible origins of this character state. The cause or causes of this change in pattern are unknown, but it is known that these characters are directly related to the habitat in which these plants develop. The genus *Acacia*, in general, contains species of secondary growth, or, species of low primary growth that develop in dry areas. Species of the myrmecophile group, on the other hand, can be found as reproductive populations under a wider range of environmental conditions of precipitation, temperature, soil quality and absolute values (as well as patterns) of disturbance (Janzen 1966).

The ability to compete with the rapidly growing vegetation of more humid habitats is a direct result of the behavior of the ants that live in the plant, because by trimming the vines and killing the intrusive branches of the adjacent trees they position the *Acacia* in a highly isolated microhabitat; also, since the meristems of this plant are intensely protected by the ants, they lack the usual structural hardness found in non-myrmecophile species and therefore have a sufficiently high vertical growth rate and branch elongation (up to 2.5 cm per day) as to make it an effective competitor in humid habitats (Janzen 1966, 1974).

This relationship enjoys from a positive feedback, because in order to keep the ant colony, the *Acacias* must procure to reduce the period during which they lose leaves, or, the season during which they diminish their production of leaf structures. Being able to inhabit humid habitats, this period is gradually reduced to almost disappearance. So, the plant can keep an increasingly large colony, increasing its competitive advantage, and produce a larger number of larger leaves and house more and more ants. In this way, the size of the colony of mutualistic ants is a function of the rate and the continuous nature of the production of leaves. Since domatia, extrafloral nectaries and Beltian bodies are all foliar structures, it seems that in this system there is a selection pressure that leads the *Acacias* to inhabit the wettest possible areas and avoid leaf loss (Janzen 1966).

In this way, it could be inferred that the increase in pinnule surface area (given by the increase in the length and width of these) during the transition from the external group to the myrmecophile group could be related to the preponderant humidity conditions in the places where these species inhabit, a condition that is also in turn influenced by the mutualistic relationship that these species establish with ants.

Regarding extrafloral nectaries, the study shows that those of the probable ancestor were circular and presented 1 to rarely 2 per petiole. Regarding the maximum length of the petiole, an increase in this measurement is observed during the transition from the external group to the myrmecophile group, that is to say, that from a range of 10 to 12.5 mm, the species went to possess petioles of 12.5 to 22.5 mm in maximum length (Table 2). Also, during this transformation, a change was observed both in the number of nectaries per petiole, and in the form of these. From presenting 1 to 2 nectaries, the species went on to show from 1 to 7 or even from 4 to 14, as in *A. chiapensis*. And from having circular nectaries, the species began to exhibit columnar nectaries (or cylindrical) and in canoe form (or in the form of a volcano, Table 2). It is possible that these modifications are closely related to each other.

Extrafloral nectaries are one of the most important adaptations to myrmecophily, in addition to being postulated as mediators in the facultative interactions that preceded the evolution of this symbiosis. Since their main function is to protect the vegetative and reproductive structures of the plant by attracting ants, and are more common in legumes with a continuous production of young leaves (McKey 1989). Thus, the foliar nectaries of myrmecophile species are larger and have a more dense nectar flow than those of non-myrmecophile species. In addition, ant-*Acacias* secrete said substance in a constitutive way and not in an induced manner. It should be mentioned that the constitutive expression of this type of indirect defenses corresponds to the apomorphic state of this character. It has been reported that some 2 m - high *Acacias* produce approximately 1 cc of nectar per day, which translates into 40 mg of a mixture of glucose and fructose. This flow is virtually the only source of sugar for the ant colony that inhabits a given individual (Heil et al. 2004; Janzen 1966, 1974). Despite the enormous energy expenditure that its production and maintenance means, the energy invested is compensated by the protection conferred by the ants against predatory and parasitic insects. In this way, nectaries tend to be more active during the most vulnerable stage of the development of leaves, flowers and fruits; and they are more common in species with a continuous production of young leaves (Janzen 1966).

In reference to the modified spinescent stipules, these represent an apomorphy of the myrmecophile group; all the species of this group possess this character state, since it is directly related to the symbiotic association between the *Acacias* and the ants. On the other hand, the species of the external group have only unmodified spinescent stipules, since the selection pressure exerted by myrmecophily does not act on them as it does on the myrmecophile group. In this way, the plesiomorphic character is that of the unmodified stipules. The analysis shows that in the ancestor, the length of the spinescent stipules varied between 46 and 55 mm (Table 2).

The domatia distinguish myrmecophytes from any other plant and correspond to one of the most important evolutionary precursors of obligate myrmecophily. In *Acacias* these structures are dilated in diameter and in length and persist in the tree for two to four years, although after one or two years they are not accepted as nesting sites by ants. The total volume of a domatium varies from 0.1 to 0.7 cc, so, usually an ant-*Acacia* contains 100 to 800 cc of internal volume; a space where all the larvae and a large part of the colony's workers live. The external surface of the domatium is resistant to abrupt temperature changes and is impermeable for at least two years, while the internal wall is absorbent and as such probably regulates the humidity of the domatia. In addition, the hardened walls of the domatia protect the ants from bird predation, especially during the dry season; whereas, the spiny shape of these structures could serve to position the *Acacias* very low in the list of food preferences of vertebrates (Janzen 1966; McKey 1989). Thus, it is not surprising that the selection pressures imposed by myrmecophily have favored those genotypes that produced spinescent stipules modified into domatia, capable of keeping a colony large enough to patrol the tree.

Regarding the evolutionary precursors of these structures, McKey (1989) postulates that relatively few plants have structures that could serve as protodomatia in facultative interactions with ants. If such pre-adaptations are rare, then the evolution of domatia is not common either. Therefore, this author suggests that myrmecophytes have evolved in some plant groups (but not in others where the ant-plant facultative interactions are equally or mostly distributed) because the former possess adequate morphological characters that could serve as pre-adaptations. In this way, myrmecophytes evolved from plants that already possessed, at least occasionally, hollow structures that resembled the original sites that ants inhabited and nested in. For example, most *Pseudomyrmex* species build small colonies in dead and hollow stems, in the soil of thick grasslands or well-developed forests (Janzen 1974).

The ants that occupied these cavities could have started consuming food from the extrafloral nectaries or from the food bodies, if they were present, but in their absence the ants could have "cultivated" homoptera in said cavities. Thus, domatia led to a greater constancy in the association with the ants than the systems where they foraged opportunistically from the plant and nested somewhere else. This constancy in the presence of the ants provided the system with improved protection and the ideal selective environment for the development of obligate interdependence with specific ants (McKey 1989).

The origin of these hollow structures in legumes could be located in areas where, in their non-myrmecophile relatives, there is a local thickening of the medulla. In the *Acacias*, the common base of the two robust spinescent stipules provided the right area for the local thickening of the medulla, which presumably dried up and produced a cavity that could have been occupied by the arboreal ants. Therefore, the presence of long spinescent stipules could have been the reason that myrmecophytes have arisen twice in the *Acacia* subgenus and not in the other genera of the Mimosoideae subfamily, so rich in facultative interactions with ants (Davidson & McKey 1993; McKey 1989).

On a more general plane, protodomatia could have been produced fortuitously in a variety of structures whose original function was not related to myrmecophily, but which derived into the ideal characters for the development of this; that is to say that the most important structures for the evolution of myrmecophily (the domatia) could constitute an example of exaptation. Many architectural features of the plants have provided the pre-adaptations necessary for the evolution of domatia, and the facultative interactions with the ants have probably acted as the selection agent that facilitated the emergence of altered shapes and sizes in these structures (McKey 1989).

Also, in relation to the presence or quality of secondary metabolites, the plesiomorphic character corresponds to the presence of S-proacacipetalin as a defense metabolite against herbivory and the apomorphic character seems to be the absence of cyanogenics in the defense of the plant. The presence of secondary metabolites is a character related to myrmecophily, because with the defense that the ants lend to the plant the presence of biochemical defenses would result in a redundant defense strategy and therefore in an unnecessary energy expenditure. Therefore, the change in this character is associated with the symbiotic relationship between *Acacias* and ants. In this way, most of the myrmecophile *Acacias* lack cyanoglucosides to combat herbivory, since the protection conferred by the ants is enough to drive away phytophagous insects (Janzen 1966, 1974).

A. *chiapensis* is the only species of the myrmecophile group in which we find S-proacacipetalin, this can be attributed to the fact that A. *chiapensis* is the basal species of the myrmecophile group and that, as mentioned before, is able to survive in the absence of mutualistic ants, although in its natural habitat it is always associated with them (Rehr et al. 1973). A. *globulifera*, on the other hand, presents the cyanoglucoside R-epiproacacipetalin, although some individuals may lack the enzyme capable of hydrolyzing it (Seigler & Ebinger 1987, 1995).

In this sense, at some point it was considered that myrmecophile plants totally lacked chemical defenses (Rehr et al. 1973), however, it has been proven that these plants retain at least some of them (Heil et al. 2002), and that the expression of chemical defenses and extrafloral nectar are not correlated in at least one plant tribe (Rudgers et al. 2004). The constitutive expression of such defenses could facilitate the transition away from mutualism (Bronstein et al. 2006).

By way of summary, it has been noted that the common ancestor of the myrmecophile group and of the external group tends to show character states with intermediate values in comparison with those presented by the ancestors of the myrmecophile group and of the A. *macracantha* group separately; or it tends to exhibit character states more similar to those observed in the external group (Table 2). The aforementioned may suggest that the characteristics that define the myrmecophile group of Neotropical *Acacias* arose and developed after the taxa of this group and the taxa of the A. *macracantha* group diverged.

Also, a key species for the understanding of the evolution of myrmecophily is A. *chiapensis*, located at the base of the myrmecophile group, which is able to survive in the absence of mutualistic ants, although in its natural habitat it is mostly associated with them. Its ability to survive in the absence of this mutualistic insect is due to the presence of cyanogenic compounds (or perhaps even alkaloids) in its leaves (Rehr et al. 1973). This species shares other characteristics with the A. *macracantha* external group, such as the number of pinnae pairs per leaf and of pinnules per pinna, the low production of Beltian bodies and a poor development of this type of food resources, as well as the scarce amount of spinescent stipules present (Janzen 1974).

## 3.3. Influence of the Paleoenvironment on the Evolution of Myrmecophytic Characters

The changes described above may also have been influenced by the paleoenvironmental context in which they were developed. It is important to consider that the common ancestor of the myrmecophile group and of the external group could have existed approximately 7.02 ± 1.95 million years ago, that is, at the end of the Miocene; whereas, the ancestor of the myrmecophile group could have lived approximately 5.44 ± 1.93 million years ago, that is, on the limit between the Miocene and the Pliocene (Gómez-Acevedo et al. 2010).

In this sense, it has been proposed that primitive versions and essentially modern versions of the current Neotropical ecosystems were generated as a result of different processes that took place during the Cretaceous and the Cenozoic. A first approximation suggests that Neotropical lowland rainforests appeared 64 million years ago; while modern versions of scrub/chaparral, savannah and grassland emerged in the middle Miocene, approximately 13 to 15 million years ago (Graham 2011).

It should be mentioned that during the middle Eocene and the early Miocene there were already primitive versions of scrub/chaparral and savannah, composed of genera such as *Acacia*, *Arctostaphyllos*, *Berberis*, *Bursera*, *Caesalpinia* and *Juniperus*. It is possible that some of these elements have come from other continents, through different routes and at different times. For example, a group from Africa and Gondwana that probably entered the American continent from the north includes *Acacia*, *Bursera*, *Caesalpinia*, *Cassia*, *Cedrela*, *Chrysophyllum*, *Ficus*, *Nectandra*, *Ocotea*, *Persea*, *Sapium*, *Sterculia* and *Terminalia* (Burnham & Graham 1999; Graham 2011).

The unique characteristics that Neotropical vegetation presents are due to three main factors: the geographical isolation of its meridional portion during most of the last 90 million years, the subsequent uplift of the Andes (which began approximately 12.9 million years ago, in the late Miocene) and the interactions between the elements coming from the north and the south, caused by the formation of the Panama land bridge. It should be noted that the most successful groups of plants that migrated from the south to the north showed a great diversification in the drier and seasonal habitats of Central America. An example of this is the trees and shrubs of the tropical forests of Mexico and Central America (Burnham & Graham 1999).

Paleobotanical and paleomastozoological evidence indicates that tropical forests predominated throughout the Cenozoic, with limited areas of savannahs and grasslands developing during the late Miocene. Locally, drier habitats with more open vegetation, began to develop during the late Miocene and early Pliocene, highlighting the first as the most important time for the development and diversification of dry communities. The early Cenozoic floras from southern Central America contain few elements of dry habitats. However, by the Miocene-Pliocene boundary, these elements were multiplied, including 11 taxa (*Poaceae*, *Acacia*, *Allophylus*, *Bursera*, *Cedrela*, *Ceiba*, *Combretum*, *Jatropha*, *Posoqueria* and *Pseudobombax*) that collectively indicate an anticipated form of tropical dry forest (Burnham & Graham 1999; Graham 2011).

Another factor that influenced the climate, especially along the coastlines, is the emergence of cold and deep waters to the sea surface, caused by the intensification of the flow of the Gulf Stream during the establishment of the Panama land bridge. This could be an explanation of the colder climates during a global warming interval. All these changes produced, periodically, appropriate local conditions for the exchange of temperate biota from open habitats at the end of the Neogene in Central America (Burnham & Graham 1999).

The Andes uplift and the consequent change in drainage patterns, as well as the formation of Panama's land bridge (both events occurred approximately 13 to 15 million years ago) caused a continuous segregation of the landscape and biota into regionally isolated populations, as well as the appearance of colder local habitats. These local conditions developed during a time interval characterized by: a significant increase in global temperature, globally arid conditions and a considerable decrease in $CO_2$ atmospheric concentration; environmental circumstances that took place approximately 6 to 8 million years ago (Burnham & Graham 1999; Coates et al. 1992; Graham 1997).

The emergence of globally dry and warm climates, but locally temperate; as well as fragmented habitats with open vegetation resulted in the emergence of ecosystems such as scrub/chaparral, savannahs and grasslands with some areas dominated by dry tropical forests. These biomes were developed during the late Miocene and early Pliocene, and were formed and enriched with species from other regions, such as Africa and South America (Burnham & Graham 1999; Graham 2011).

It is in these preponderantly warm-humid, although limited, areas where the vegetable characters that facilitate myrmecophily could have developed. That is, the warm-humid conditions of the environment could have favored the evolution of characters such as: an increased and continuous production of leaves and the retention of these by most individuals, a high horizontal and vertical growth rate, and the development of generally larger pinnules. This guaranteed a constant flow of food (through Beltian bodies and extrafloral nectaries) and the permanent provision of ideal nesting sites (domatia) for the ant colony. It is worth mentioning that these characters evolved not only from the paleoenvironmental context, but rather because of the selection pressure that these insects imposed on them, and the paleoenvironmental circumstances simply favored the development of characters that facilitated this association.

In this way, favored both by the restricted conditions of humidity, and by the positive feedback that existed between these conditions and myrmecophily; the opportunistic and/or antagonistic relationships between the *Acacias* and the ants were progressively transformed into obligate relationships. The selection pressure that the interaction with the ants exerted on the *Acacias* promoted the evolution of the highly specialized characters that distinguish the *Acacia-Pseudomyrmex* mutualistic system.

The current distribution of Neotropical myrmecophile *Acacias* is one of the evidences that could sustain this observation, because unlike non-myrmecophile species, the former are not able to survive in dry environments. The fragmentation of the landscape could also have participated in the evolution of myrmecophilic characters, since the segregation of biota into regionally isolated populations could result both in processes of allopatric speciation, as well as in a more constant and close interaction between the *Acacias* and the ants, at the decrease of the distribution area of both participants.

## CONCLUSION

The reconstruction based on the maximum parsimony analysis is congruent with that obtained through the maximum likelihood analysis, however, the results achieved through the second method are clearer than those achieved by the first.

The common ancestor of the myrmecophile group and the *A. macracantha* group probably presented intermediate character states between both groups; nevertheless, some of these states, such as maximum height and the shape of the extrafloral nectaries, resembled more those present in the external group.

In the transition from the external group to the myrmecophile group: the leaves lengthened in a first step and then became smaller compared to their original size, the number of pinnae pairs per leaf and of pinnules per pinna decreased, the length and width of the pinnules increased, the maximum length of the petiole increased, the number of extrafloral nectaries per petiole also increased and the shape of the nectaries changed from circular to columnar (or cylindrical) and canoe shaped (or volcano), the spinescent stipules were modified into domatia and grew in length, and finally the species lost their cyanoglucosides against herbivory. In this way, the selection pressure that myrmecophily exerts on the interacting species is reflected in a systematic set of their attributes.

The characteristics that identify the Neotropical myrmecophile *Acacias* were manifested and developed after the lineages of this and the *A. macracantha* group diverged, that is, approximately 5.44 million years ago, at the end of the Miocene. In this way, the vegetal characters that facilitated myrmecophily in the *Acacia-Pseudomyrmex* mutualistic system may have evolved as they were favored by the limited warm-humid conditions that predominated in the ancient Neotropical environments. Finally, for a more detailed description and discussion of this study, please consult Córdoba-de León (2017).

# REFERENCES

Bronstein, J. L. (1994). Our Current Understanding of Mutualism. *The Quarterly Review of Biology, 69 (1)*, 31–51. https://doi.org/10.1086/418432.

Bronstein, J. L., Alarcon, R., & Geber, M. (2006). The evolution of plant-insect mutualisms. *The New Phytologist, 172 (3)*, 412–428. https://doi.org/10.1111/j.1469-8137.2006.01864.x.

Burnham, R. J., & Graham, A. (1999). The History of Neotropical Vegetation: New Developments and Status. *Annals of the Missouri Botanical Garden, 86 (2)*, 546–589. https://doi.org/10.2307/2666185.

Chenuil, A., & Mckey, D. (1996). Molecular phylogenetic study of a myrmecophyte symbiosis: did *Leonardoxa*/Ant associations diversify via cospeciation?. *Molecular Phylogenetics and Evolution, 6 (2)*, 270-286. https://doi.org/10.1006/mpev.1996.0076.

Coates, A. G., Jackson, J. B. C., Collins, L. S., Cronin, T. M., Dowsett, H. J., Bybell, L. M., Jung, P., & Obando, J. A. (1992). Closure of the Isthmus of Panama: The near-shore marine record of Costa Rica and western Panama. *GSA Bulletin, 104 (7)*, 814–828. https://doi.org/10.1130/0016-7606(1992)104<0814:COTIOP> 2.3.CO;2.

Collins, T. M., Wimberger, P. H., & Naylor, G. J. P. (1994). Compositional Bias, Character-State Bias, and Character-State Reconstruction Using

Parsimony. *Systematic Biology*, *43 (4)*, 482–496. https://doi.org/10.2307/2413547.

Córdoba-de León, J. A. (2017). *Reconstrucción de los estados de carácter ancestrales de las Acacias mirmecófilas*. Universidad Nacional Autónoma de México.[Ancestral character state reconstruction of ant-Acacias. National Autonomous University of Mexico].

Cunha, A. F., Genzano, G. N., & Marques, A. C. (2015). Reassessment of Morphological Diagnostic Characters and Species Boundaries Requires Taxonomical Changes for the Genus *Orthopyxis* L. Agassiz, 1862 (Campanulariidae, Hydrozoa) and Some Related Campanulariids. *PLOS ONE*, *10*(2), e0117553. Retrieved from https://doi.org/10.1371/journal.pone.0117553.

Cunningham, C. W., Omland, K. E., & Oakley, T. H. (1998). Reconstructing ancestral character states: a critical reappraisal. *Trends in Ecology and Evolution*, *13 (9)*, 361–366.

Davidson, D. W., & McKey, D. (1993). The evolutionary ecology of symbiotic ant-plant relationships. *Journal of Hymenoptera Research*, *2*, 13–83.

de Oliveira, L. L., Calazans, L. S. B., de Morais, E. B., Mayo, S. J., Schrago, C. G., & Sakuragui, C. M. (2014). Floral evolution of *Philodendron* subgenus *Meconostigma* (Araceae). *PloS One*, *9 (2)*, e89701. https://doi.org/10.1371/journal.pone.0089701.

Fiala, B., Jakob, A., Maschwitz, U., and Linsenmair, K. E. (1999). Diversity, evolutionary specialization and geographic distribution of a mutualistic ant-plant complex: Macaranga and Crematogaster in South East Asia. *Biological Journal of the Linnean Society*, *66 (3)*, 305–331. https://doi.org/10.1111/j.1095-8312.1999.tb01893.x.

Finarelli, J. A., & Flynn, J. J. (2006). Ancestral state reconstruction of body size in the Caniformia (Carnivora, Mammalia): the effects of incorporating data from the fossil record. *Systematic Biology*, *55*, 301–313. https://doi.org/10.1080/10635150500541698.

Givnish, T. J., Pires, J. C., Graham, S. W., McPherson, M. A., Prince, L. M., Patterson, T. B., Rai, H. S., Roalson, E. H., Evans, T. M., Han, W. J., Milam, K. C., Meerow, A. W., Molvray, M., Kores, P. J., O'Brien, H.

E., Hall, J. C., Krees, W. J., & Sytsma, K. J. (2005). Repeated evolution of net venation and fleshy fruits among monocots in shaded habitats confirms a priori predictions: evidence from an ndhF phylogeny. *Proceedings of the Royal Society B, 272*, 1481 LP-1490. DOI: 10.1098/rspb.2005.3067.

Gómez-Acevedo, S., Rico-Arce, L., Delgado-Salinas, A., Magallon, S., & Eguiarte, L. E. (2010). Neotropical mutualism between *Acacia* and *Pseudomyrmex*: phylogeny and divergence times. *Molecular Phylogenetics and Evolution, 56*, 393–408.

Graham, A. (1997). Neotropical Plant Dynamics During the Cenozoic-Diversification, and the Ordering of Evolutionary and Speciation Processes. *Systematic Botany, 22 (1)*, 139–150. https://doi.org/10.2307/2419682.

Graham, A. (2011). The age and diversification of terrestrial New World ecosystems through Cretaceous and Cenozoic time. *American Journal of Botany, 98*, 336–351. https://doi.org/10.3732/ajb.1000353.

Harvey, P. H., & Pagel, M. D. (1991). *The comparative method in evolutionary biology*. Oxford, UK: Oxford University Press.

Heil, M. (2008). Indirect defence via tritrophic interactions. *New Phytologist, 178*, 41–61. https://doi.org/10.1111/j.1469-8137.2007.02330.x.

Heil, M., Baumann, B., Krüger, R., & Linsenmair, K. E. (2004). Main nutrient compounds in food bodies of Mexican *Acacia* ant-plants. *Chemoecology, 14*, 45–52. https://doi.org/10.1007/s00049-003-0257-x.

Heil, M., Delsinne, T., Hilpert, A., Schürkens, S., Andary, C., Linsenmair, K. E., Sousa, S. M. and McKey, D. (2002). Reduced chemical defence in ant-plants? A critical re-evaluation of a widely accepted hypothesis. *Oikos, 99 (3)*, 457–468. https://doi.org/10.1034/j.1600-0706.2002.11954.x.

Heil, M., & McKey, D. (2003). Protective Ant-Plant Interactions as Model Systems in Ecological and Evolutionary Research. *Annual Review of Ecology, Evolution, and Systematics, 34 (1)*, 425–553. https://doi.org/10.1146/annurev.ecolsys.34.011802.132410.

Horn, J. W., Fisher, J. B., Tomlinson, P. B., Lewis, C. E., & Laubengayer, K. (2009). Evolution of lamina anatomy in the palm family (Arecaceae). *American Journal of Botany*, 96 (8), 1462–1486. https://doi.org/10.3732/ajb.0800396.

Howe, H. F., & Westley, L. C. (1989). *Ecological relationships of plants and animals*. New York: Oxford University Press.

Janzen, D. (1966). Coevolution of mutualism between ants and *Acacias* in Central America. *Evolution*, 20 (3), 249–275. https://doi.org/10.1111/j.1558-5646.1966.tb03364.x.

Janzen, D. (1974). *Swollen-Thorn Acacias of Central America. Smithsonian Contributions to Botany* (Vol. 13). https://doi.org/10.5479/si.0081024X.13.

Lewis, G. P., Schrire, B., Mackinder, B., & Lock., M. (2005). *Legumes of the World*. Royal Botanic Gardens, Kew.

Maddison, W. P., & Maddison., D. R. (2015). M*esquite: a modular system for evolutionary analysis*. Version 3.04.

Mayer, V. E., Frederickson, M. E., McKey, D., & Blatrix, R. (2014). Current issues in the evolutionary ecology of ant-plant symbioses. *New Phytologist*, 202 (3), 749–764. https://doi.org/10.1111/nph.12690.

McKey, D. (1989). Interactions between ants and leguminous plants. In C. H. Stirton and J. L. Zarucchi (Eds.), *Advances in Legume Biology* (pp. 673–718). St. Louis, Missouri: Monograph in Systematic Botany of the Missouri Botanical Garden.

Miller, J. T. & Seigler D. (2012). Evolutionary and taxonomic relationships of *Acacia s.l.* (Leguminosae: Mimosoideae). *Australian Systematic Botany*, 25, 217-224.

Mooers, A. O., Vamosi, S. M., Schluter, D., & Pagel, A. E. M. (1999). Using Phylogenies to Test Macroevolutionary Hypotheses of Trait Evolution in Cranes (Gruinae). *The American Naturalist*, 154 (2), 249–259. https://doi.org/10.1086/303226.

Nei, M., & Kumar., S. (2000). *Molecular evolution and phylogenetics*. New York, USA: Oxford University Press.

Nepokroeff, M., Sytsma, K. J., Wagner, W. L., & Zimmer, E. A. (2003). Reconstructing ancestral patterns of colonization and dispersal in the

Hawaiian understory tree genus Psychotria (Rubiaceae): a comparison of parsimony and likelihood approaches. *Systematic Biology, 52 (6)*, 820–838.

Nurk, N. M., Madrinan, S., Carine, M. A., Chase, M. W., & Blattner, F. R. (2013). Molecular phylogenetics and morphological evolution of St. John's wort (Hypericum; Hypericaceae). *Molecular Phylogenetics and Evolution, 66 (1)*, 1–16. https://doi.org/10.1016/j.ympev.2012.08.022.

Omland, K. E. (1999). The Assumptions and Challenges of Ancestral State Reconstructions. *Systematic Biology, 48 (3)*, 604–611.

Pagel, M. (1999a). Inferring the historical patterns of biological evolution. *Nature, 401*, 877. Retrieved from http://dx.doi.org/10.1038/44766.

Pagel, M. (1999b). The Maximum Likelihood Approach to Reconstructing Ancestral Character States of Discrete Characters on Phylogenies. *Systematic Biology, 48 (3)*, 612–622. Retrieved from http://dx.doi.org/10.1080/106351599260184.

Pagel, M., Meade, A., & Barker, D. (2004). Bayesian estimation of ancestral character states on phylogenies. *Systematic Biology, 53 (5)*, 673–684.

Rehr, S. S., Feeny, P. P., & Janzen, D. H. (1973). Chemical Defence in Central American Non-Ant-*Acacias*. *Journal of Animal Ecology, 42 (2)*, 405–416.

Revell, L. J. (2014). Ancestral character estimation under the threshold model from quantitative genetics. *Evolution, 68 (3)*, 743–759. https://doi.org/10.1111/evo.12300.

Ricklefs, R. E. (2007). Estimating diversification rates from phylogenetic information. *Trends in Ecology and Evolution, 22 (11)*, 601–610.

Rico-Arce, L. (2003). *Geographical patterns in neotropical Acacia (Leguminosae: Mimosoideae). Australian Systematic Botany* (Vol. 16). https://doi.org/10.1071/SB01028.

Rico-Gray, V., & Oliveira, P. S. (2007). *The ecology and evolution of ant-plant interactions*. Chicago, Illinois, USA: The University of Chicago Press.

Ronquist, F. (2004). Bayesian inference of character evolution. *Trends in Ecology and Evolution, 19 (9)*, 475–481. https://doi.org/10.1016/j.tree.2004.07.002.

Royer-Carenzi, M., Pontarotti, P., & Didier, G. (2013). Choosing the best ancestral character state reconstruction method. *Mathematical Biosciences, 242 (1)*, 95–109. https://doi.org/10.1016/j.mbs.2012.12.003.

Rudgers, J. A., Strauss, S. Y., & Wendel, J. F. (2004). Trade-offs among anti-herbivore resistance traits: insights from Gossypieae (Malvaceae). *American Journal of Botany, 91 (6)*, 871–880. https://doi.org/10.3732/ajb.91.6.871.

Schäffer, S., Koblmüller, S., Pfingstl, T., Sturmbauer, C., & Krisper, G. (2010). Ancestral state reconstruction reveals multiple independent evolution of diagnostic morphological characters in the "Higher Oribatida" (Acari), conflicting with current classification schemes. *BMC Evolutionary Biology, 10*, 246. https://doi.org/10.1186/1471-2148-10-246.

Schultz, T. R., Cocroft, R. B., & Churchill, G. A. (1996). The Reconstruction Of Ancestral Character States. *Evolution, 50 (2)*, 504–511. https://doi.org/10.1111/j.1558-5646.1996.tb03863.x.

Seigler, D. S., & Ebinger, J. E. (1987). Cyanogenic Glycosides in Ant-*Acacias* of Mexico and Central America. *The Southwestern Naturalist, 32 (4)*, 499–503. https://doi.org/10.2307/3671484.

Seigler, D. S., & Ebinger, J. E. (1995). Taxonomic Revision of the Ant-*Acacias* (Fabaceae, Mimosoideae, *Acacia*, Series *Gummiferae*) of the New World. *Annals of the Missouri Botanical Garden, 82*(1), 117–138. https://doi.org/10.2307/2399983.

Simpson, M. G. (2010). *Plant Systematics* (2º ed.). Massachusetts: Academic Press.

Soltis, D. E., Mort, M. E., Latvis, M., Mavrodiev, E. V, O'Meara, B. C., Soltis, P. S., Burleigh, J. G., Rubio de Casas, R. (2013). Phylogenetic relationships and character evolution analysis of Saxifragales using a supermatrix approach. *American Journal of Botany*, *100 (5)*, 916–929. https://doi.org/10.3732/ajb.1300044.

Vanderpoorten, A., & Goffinet, B. (2006). Mapping uncertainty and phylogenetic uncertainty in ancestral character state reconstruction: an example in the moss genus Brachytheciastrum. *Systematic Biology*, *55 (6)*, 957–971.

Webster, N. B., Van Dooren, T. J. M., & Schilthuizen, M. (2012). Phylogenetic reconstruction and shell evolution of the Diplommatinidae (Gastropoda: Caenogastropoda). *Molecular Phylogenetics and Evolution*, *63 (3)*, 625–638.

Wiley, E. O., & Lieberman, B. S. (2011). *Phylogenetics: the theory of phylogenetic systematics* (2º ed.). New Jersey, USA: Wiley-Blackwell.

Wu, Z.-Y., Milne, R. I., Chen, C.-J., Liu, J., Wang, H., & Li, D.-Z. (2015). Ancestral State Reconstruction Reveals Rampant Homoplasy of Diagnostic Morphological Characters in Urticaceae, Conflicting with Current Classification Schemes. *PLOS ONE*, *10 (11)*, e0141821. Retrieved from https://doi.org/10.1371/journal.pone.0141821.

## BIOGRAPHICAL SKETCHES

### *Jessica Admin Córdoba de León*

**Education:** Bachelor Degree in Biology by the National Autonomous University of Mexico

**Research and Professional Experience:** Evolutionary Biology, Interactions ant-plant.

**Honors:** Erasmus Mundus Master Programme in Evolutionary Biology (Cohort 2017-2019).

## Sandra Luz Gómez Acevedo

**Affiliation:** Unidad de Morfología y Función. Facultad de Estudios Superiores Iztacala, Universidad Nacional Autónoma de México. Tlalnepantla, Estado de México, México.

**Education:** PhD in Biological Sciences, Master Degree in Systematics, Bachelor Degree in Biology (all by the National Autonomous University of Mexico)

**Research and Professional Experience:** Evolutionary Biology, Floral development, Interactions ant-plant.

In: *Acacia*
Editor: Aide Matheson

ISBN: 978-1-53614-237-2
© 2018 Nova Science Publishers, Inc.

Chapter 6

# *ACACIA* POLLEN: AN IMPORTANT CAUSE OF POLLINOSIS IN TROPICAL AND SUBTROPICAL REGIONS

*Mohammad-Ali Assarehzadegan*[*], *PhD*
Immunology Department, School of Medicine
and Immunology Research Center,
Iran University of Medical Sciences (IUMS), Tehran, Iran

## ABSTRACT

Pollinosis, also known as pollen allergy, hay fever or seasonal allergic rhinitis is one the most common respiratory disorders throughout the world. The main cause of pollinosis is pollen of different species of trees. Different species of *Acacia* are common throughout tropical and subtropical regions of Asia, Africa, Australia, and America with hot and humid climates, where it is planted as a shade and/or ornamental tree or for binding sand. Generally, pollens from different members of the Fabaceae family, particularly *Acacia* Spp. and Prosopis Spp., have been reported as an important source of pollinosis in the United States, European countries,

---

[*] Corresponding Author Email: assarehma@gmail.com & assareh.ma@iums.ac.ir.

and Asia. Moreover, the inhalation of *Acacia* pollen is one of the main causes of respiratory allergic diseases in semiarid countries such as Iran, Saudi Arabia, and the United Arab Emirates, where the frequency of sensitization ranges from 25% to 48%.

The crude extracts from trees pollen are usually composed of several protein and glycoprotein components. The protein molecules are rapidly diffused after contact with mucus membranes of humans. The recognition of allergenic components of pollens is essential for component-resolved diagnosis, the design of patient-specific immunotherapy, and the explanation of sensitization mechanisms to various allergens.

Immunochemical analysis of the protein profile of *Acacia farnesiana* (*Vachellia farnesiana*) pollen indicated several components ranging from 12 to 105 kilodalton (kDa). Moreover, the 15, 45, 50, 66 and 85 kDa proteins are recognized as dominant IgE-binding components of *Acacia* pollen. Despite a high prevalence of sensitization to *Acacia* pollen in semiarid countries, there is little information about the molecular characterization of *A. farnesiana* pollen allergens. In one recent report, a new allergen from *Acacia* pollen was introduced for the first time. This allergen was named Aca f 1 in accordance with the International Union of Immunological Societies (IUIS) Allergen Nomenclature Subcommittee. Aca f 1 is a member of the Ole e 1-like protein family. Aca f 2, other new allergen from *A. farnesiana* pollen, is an allergenic member of the profilin family. Profilins are 12- to 15-kDa monomeric proteins belonging to a ubiquitous family of actin-binding proteins. They are prominent allergens in the pollens of trees, grasses, and weeds, and many fruits and vegetables.

Cross-reactivity among allergen sources occurs when the allergenic proteins from one source of allergen are similar to the allergenic proteins in another source of allergen. Cross-reactivity among *Acacia* pollen components with *Lolium perenne* pollen allergens has been described. Moreover the results of earlier studies indicated a significant IgE cross-reactivity between *Acacia* and *P. juliflora* (Mesquite) pollens. In general, the knowledge of pollen cross-reactivity is crucial to diagnostics as well as formulation of immunotherapy vaccines. Furthermore a high degree of sequence similarity (90%) was detected between Aca f 1 and Sal k 5 (Allergenic profilin of *Salsola kali* pollen). Besides the results of amino acid sequence similarity analysis revealed that Aca f 2 has a high degree of identity with the selected profilins from the most common allergenic regional plants, particularly Pro j 2 (*P. juliflora* profilin) (96%).

**Keywords:** *Acacia*, pollinosis, allergy, prevalence, Aca f 1, Aca f 2

# INTRODUCTION

Pollinosis, also known as pollen allergy, hay fever or seasonal allergic rhinitis is one the most common respiratory disorders throughout the world. The main cause of pollinosis is pollen of different species of trees. Generally pollen of plants contains numerous protein molecules nevertheless only a small part of these molecules able to induce allergic reactions in humans [1]. Following the contact of allergenic pollen with the mucus membranes of susceptible individuals, symptoms of pollinosis typically develop including sneezing, runny and itchy nose and eyes, conjunctivitis and sometimes bronchial asthma [1, 2].

Prevalence of pollinosis has been expected up to 40% in some regions in the world [3]. The occurrence of pollinosis varies among different countries, as well as different regions within a country, which could be associated to the variety of allergens existing in those areas [4-7]. Based on earlier studies, pollens of different species of plants play an important role in pathogenesis of pollinosis. Besides weeds and grasses pollens, trees particularly wind-pollinated ones with heavy pollen productions are important sources of allergenic pollens that causes hay fever or seasonal allergic rhinitis. Some of the common allergenic trees are listed in Table 1.

**Table 1. Some of the common allergenic trees in the world**

| Family | Botanic name | Common name |
|---|---|---|
| Fabaceae | *Prosopis juliflora* | Mesquite |
|  | *Acacia farnesiana* | *Acacia* |
|  | *Albizia lebbeck* |  |
| Platanaceae | *Platanus orientalis* | Sycamore |
| Oleaceae | *Fraxinus excelsior* | Ash |
|  | *Olea europaea* | Olive tree |
| Salicaceae | *Populus alba* | White poplar |
| Aceraceae | *Acer cappadocicum* | Maple tree |
| Simaroubaceae | *Ailanthus altissima* | Tree of heaven |
| Cupressaceae | *Cupressus sempervirens* | Cypress |
| Myrtaceae | *Eucalyptus camaldulensis* | Eucalyptus |
| Pinaceae | *Pinus eldarica* | Pine |
| Cupressaceae | *Juniperus communis* | Juniper |

## *Acacia* Pollinosis

Different species of *Acacia* are common throughout tropical and subtropical regions of Asia, Africa, Australia, and America with hot and humid climates, where it is planted as a shade and/or ornamental tree or for binding sand [4, 18, 19]. Generally, pollens from the Fabaceae family have been described as an important source of pollinosis in the United States, European countries, and different regions of Asia [4, 18-20] (Table 2). Moreover, the inhalation of *Acacia* pollen is one of the main causes of respiratory allergic diseases in semiarid countries such as Iran, Saudi Arabia, and the United Arab Emirates, where the incidence of sensitization to pollen of *Acacia* is range between 25% and 48% [4, 7, 19, 21-23] (Table 2).

**Table 2. Prevalence of *Acacia* pollen allergy in different regions of the world**

| Author (Publication year) Ref | Country | Prevalence | Patients |
|---|---|---|---|
| Kung Sh.Ju et al. (2013) [28] | Botswana | 21% | Atopic |
| Bousquet J. et al. (1984)[29] | France (Montpellier area) | 7% | Pollen allergic |
| Baratawidjaja et al. (1999) [30] | Indonesia | 12% | Allergic rhinitis and/or Asthma |
| Assarehzadegan M.-A. et al. (2013) [4] | Iran (Tropical regions) | 48% | Allergic rhinitis |
| Assarehzadegan M.-A. et al. (2013) [21] | Iran (Tropical regions) | 45% | Asthma |
| Fereidouni M. et al. (2009) [7] | Iran (Northern regions) | 29 | Allergic rhinitis |
| Sam CK. et al.(1998) [31] | Malaysia | 21.5% | Asthma |
| Liam CK. et al. (2002) [32] | Malaysia | 8% | Asthma |
| Bener A. et al. (2002) [22] | United Arab Emirates (UAE) | 25.6% | Asthma |
| Suliaman F.A. et al. (1997) [23] | Saudi Arabia | 29% | Atopic |

It worthy to note that in previous studies different species of *Acacia* has been reported to the cause of occupational asthma [24, 25]. The results indicated that gum Arabic, also known as *Acacia* gum, may cause occupational allergic rhinitis and asthma along with urticaria or eczema symptoms in the candy factory workers when there was an increase in the contact and exposure to it [24-26]. Moreover, it has been reported occupational airborne contact dermatitis after contact to *Acacia melanoxylon* wood [27].

## Physicochemical and Immunological Characterization of *Acacia* Pollen

One approach to expand insight into the pathogenesis of allergic disorders is the characterization of allergens. The use of molecular biology methods during the last decades have certainly improved the knowledge about the structure and biological function of aeroallergens such as trees pollen grains [33]. The crude extracts from trees pollen are usually composed of several proteins and glycoprotein components. Up to now, the characterization of trees pollen allergens indicated low-molecular-weight proteins or/and glycoproteins molecules with relatively high solubility, stability and glycosylation properties. These proteins are rapidly diffused after contact with mucus membranes of humans. The recognition of allergenic components of pollens is essential for component-resolved diagnosis, the design of patient-specific immunotherapy, and the explanation of sensitization mechanisms to various allergens [19, 34, 35]. The combination of data from molecular and structural evaluation of IgE-binding components from pollen extracts is critical for component-resolved diagnosis, specific immunotherapy, and the elucidation of mechanisms underlying sensitization to various allergens [34, 36].

Immunochemical analysis of the protein profile of *Acacia farnesiana* (*Vachellia farnesiana*) pollen grain indicated several components range from 12 to 105 kilodalton (kDa). Moreover, the 15, 45, 50, 66 and 85 kDa proteins are documented as main IgE-binding components of *Acacia* pollen grain [19, 37, 38].

In our previous study [19], the protein composition of *A. farnesiana* pollen extract was analyzed by Coomassie Brilliant Blue staining (Figure 1A). The reducing SDS-PAGE separation of the pollen extract showed several resolved proteins with molecular weights in the range of approximately 12 to 85 kDa. The most prominent bands had MWs of about 50, 66, and 85kDa. Other predominant bands were identified at 12, 15, 20, 23, 28, 37, and 45 kDa. Non-reducing SDS-PAGE of *Acacia* pollen extract showed eight prominent proteins with estimated molecular weights between 15 and 66 kDa (Figure 1A).

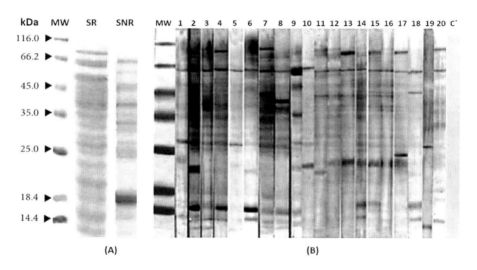

Figure 1. (A). Coomassie Brilliant Blue stained SDS-PAGE of the crude extract of *A. farnesiana* pollen in reducing and non-reducing conditions on 12.5% polyacrylamide gel. Lane MW, molecular weight marker (Thermo Fisher scientific, USA); SR, reducing condition; SNR, non-reducing condition. (B). Immunoblotting of *Acacia* pollen extract (with reducing SDS-PAGE). Each strip was first blotted with *Acacia* pollen extract. All strips were then incubated with allergic patients' sera and detected for IgE reactive protein fractions. C⁻, negative control.

For the evaluation of IgE-biding reactivity of *Acacia* pollen proteins the separated protein bands of the electrophoresis of *Acacia* pollen extract was determined by immunoblotting assays. Specific IgE binding proteins probed with sera from twenty *Acacia* allergic patients are shown in Figure 1B. The results showed several IgE reactive bands ranging from about 12 to 85 kDa. Figure 1B shows the apparent MW of each protein band and the prevalence of each one among all twenty allergic patients. The most frequent IgE reactive protein bands among the patients' sera were 66 and 85 kDa. However, there were other IgE reactive proteins among patients' sera with molecular weights 15, 20, 23, 28, 39, 45, and 50 kDa (Figure 1B).

## Identification of Novel Allergens from *Acacia* Pollen

Genes and cDNA sequences encoding allergenic proteins (allergens) from trees pollen have been cloned and sequenced. Cloning and recombinant expression of these allergens has allowed determination of their molecular and immunological properties including amino acid sequence, secondary and three-dimensional structures and IgE-binding sites (linear and conformational epitopes). Moreover, cloning of pollen allergens and identification of molecular structure of allergenic pollens and other allergens from different sources can revealed sequence and structural homology among these allergenic proteins and enzymes. By using the homology data we can predict the function and potential cross-reactivity of the new allergens [1, 39].

## Aca f 1

Despite a high prevalence of sensitization to *Acacia* pollen in semiarid countries, there is little information about the molecular characterization of *A. farnesiana* pollen allergens. In our recent report, a new allergen from *Acacia* pollen was introduced for the first time [40]. This allergen was named Aca f 1 in accordance with the International Union of Immunological Societies (IUIS) Allergen Nomenclature Subcommittee (http://www.allergen.org/). Aca f 1 is a member of the Ole e 1-like protein family. The first allergenic member of Ole e 1-like family was isolated from *Olea europaea* pollen as the major allergen of olive pollen (Ole e 1) [41]. In earlier studies, similar allergens were also identified in other plants such as *Fraxinus excelsior* (Fra e 1) [42], *Ligustrum vulgare* (Lig v 1) [43] *Syringa vulgaris* (Syr v 1) [44], *Salsola kali* (Sal k 5) [45] and *Chenopodium album* (Che a 1) [46]. The IgE-reactivity to some allergenic members of Ole e 1-like protein family has been demonstrated between 30% and 77% in allergic patients in different regions where it was considered as a major allergen in pollen of the allergenic plants such as *O. europaea* (Ole e 1), *F. excelsior* (Fra e 1), *C. album* (Che a 1), *S. kali* (Sal k 5) and *Ligustrum vulgare* (Lig v 1) [43, 45-48].

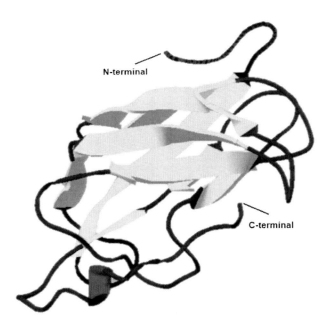

Figure 2. Homology Model of three-dimensional structure of Aca f 1 *Acacia* Allergen. Ribbon depiction of the structure of Aca f 1 modeled using I-TASSER software (https://zhanglab.ccmb.med.umich.edu/I-TASSER/).

The open reading frame (ORF) of Aca f 1 gene comprises 453 bases encoding Aca f 1 polypeptide of 150 amino acids, with six cysteine residues. This ORF is a 16.5 kDa protein that reveals molecular characteristics similar to known members of the plant Ole e 1-like protein family (ranging from 16 to 19 kDa) [42, 45, 47]. Furthermore, Aca f 1, like Sal k 5, Cro s 1 and Che a 1 has a conserved sequence for potential N-glycosylation site in the same point of the polypeptide chain (Asn-Ile/Leu-Thr-Ala), which is essentially occupied by a glycan in these proteins.

The results of immunoblotting experiments on *Acacia* pollen extract using pooled sera from patients as well, indicated an IgE-binding protein with molecular weight about 17 kDa. A three-dimensional homology model of Aca f 1 was generated by the Internet server I-TASSER (https://zhang lab.ccmb.med.umich.edu/I-TASSER/) (Figure 2).

## Aca f 2

Aca f 2, other new allergen from *A. farnesiana* pollen, is an allergenic member of the profilin family. Profilins are 12- to 15-kDa monomeric proteins belonging to a ubiquitous family of actin-binding proteins. They are prominent allergens in the pollens of trees, grasses, and weeds, and many fruits and vegetables [49-52], but these allergens are usually under-represented in natural extracts used for diagnostic, therapeutic, and experimental purposes.

Generally profilins from various sources have indicated high sequence identities and may play roles in IgE-cross reactivity in allergic patients. In earlier studies, several allergens from this family, such as Sal k 4, Ama r 2, Che a 2 and Ole e 2, have already been identified [49, 52-54]. Profilin from *A. farnesiana* pollen was cloned by a PCR strategy using degenerated primers based on the codons of conserved amino acid sequences from various plant profilins. The open reading frame (ORF) of *A. farnesiana* profilin contained 399 bases encoding a 14.2 kDa protein that correlates with the molecular characteristics of known plant profilins. The findings of immunoblotting experiments of *A. farnesiana* pollens extract using pooled sera from the patients was also indicated an IgE-binding protein band with an estimated MW of 15 kDa. A three-dimensional homology model of Aca f 2 was generated by the Internet server homology-modeling server (https://swissmodel.expasy.org/) (Figure 3). Until now, different molecular weights (MWs) from profilins of different plant sources have been described, such as 14.2 to 14.6 kDa in three members of the Amaranthaceae family (Sal k 4, Che a 2, Ama r 2), 15 kDa in *Olea europaea* pollen (Ole e 2), and 14.0 kDa in *Artemisia vulgaris* (Art v 4) [49, 52-55]. These variations may be explained by diversities in some amino acid residues, levels of glycosylation or differences in the methods of measuring MWs [52].

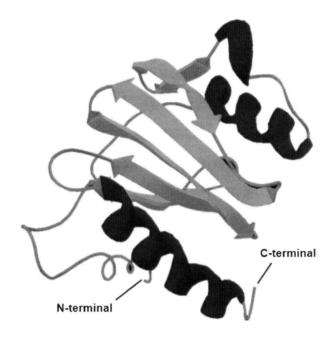

Figure 3. Homology model three-dimensional structure of Aca f 2 *Acacia* allergen. Ribbon depiction of the structure of Aca f 2 modeled according to the three-dimensional structure of the profilin of *Hevea brasiliensis*, Hev b 8 (PDB no. 1g5u.1.A).

Recently, cross-reactivity of *Acacia* pollen allergens with other allergenic plants has been described [19, 56]. In tropical areas, the importance of *Acacia* and the most allergenic members of the Amaranthaceae family (*S. kali*, *A. retroflexus*, *C. album*) and Quercus genus pollens have been reported as causing respiratory allergic disorders [21, 57-59].

The results of amino acid sequence identity evaluation indicated that Aca f 2 has a high degree of identity with some profilins from the most common allergenic local plants, predominantly Pro j 2 (*P. juliflora* profilin) (95%). This was expected because *Acacia* and *P. juliflora* both belong to the Fabaceae family. Generally, this high degree of homology may predict the similar tertiary structure.

## Cross-Reaction among Allergenic Proteins of *Acacia* Pollen Extract and Other Allergenic Proteins

Cross-reactivity among allergen sources occurs when the allergenic proteins from one source of allergen are similar to the allergenic proteins in another source of allergen. In other words, the cross-reactivity is an immune response of an IgE antibody that can recognize similar other allergenic molecules, often resulting from conservation of IgE-binding site (epitope) regions. It worthy to note that cross-reactivity always arises among allergenic molecules in thoroughly related species or sometimes among molecules from different species with related function that belongs to the same protein family.

Cross-reactivity among *Acacia* pollen components with *Lolium perenne* pollen allergens has been described [8]. The results of immunoblotting inhibition revealed that the IgE binding reactivity of the allergenic proteins with molecular weights 12, 20, 39, 45 and 66 kDa from the *A. farnesiana* pollen extract was partly inhibited by *Prosopis juliflora*, and several members of Amaranthaceae family (*Salsola kali*, *Amaranthus retroflexus*, *Chenopodium album* and *Kochia scoparia*) pollen extracts [19, 39]. The findings of these studies indicated a significant IgE cross-reactivity or in other words suggested a close allergenic relationship between *Acacia* and *P. juliflora* (Mesquite) pollens. In general, the knowledge of pollen cross-reactivity is crucial to diagnostics as well as formulation of immunotherapy vaccines. Cross-reactivity among pollens belonging to the same genus and/or diverse genera has been revealed earlier [60, 61]. Several cross-reactive protein molecules have been genetically engineered and are found to have possible for use in immunotherapy [62, 63].

## Cross-Reaction of Aca f 1 with Other Members of Ole-Like Protein Family

In previous study to investigate sequence similarity between Aca f 1 and the well-known allergenic Ole e 1-like protein in the protein database, the presumed amino acid sequences of these allergenic proteins were compared (Figure. 4) [40]. A high degree of sequence similarity (90%) was detected between Aca f 1 and Sal k 5 (Allergenic profilin of *Salsola kali* pollen) (Table 3) [40].

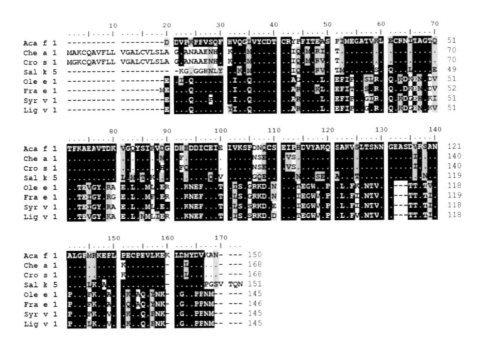

Figure 4. Comparison of the Aca f 1 (*Acacia farnesiana* Ole e 1-like protein) amino acid sequence with allergenic Ole e 1-like protein from other plants: *Chenopodium album* (Che a 1, G8LGR0.1), *Crocus sativus* (Cro s 1, XP004143635.1), *Salsola kali* (Sal k 5, ADK22842.1), *Olea europaea* (Ole e 1, P19963.2), *Fraxinus excelsior* (Fra e 1, AAQ83588.1), *Syringa vulgaris* (Syr v 1, S43243), and *Ligustrum vulgare* (Lig v 1, O82015.2). The amino acid sequence similarity of Aca f 1 (KR870435) to other members of the Ole e 1-like family are indicated in Table 3.

**Table 3. Percentage of similarity among Aca f 1 and some allergenic Ole e 1-like proteins**

| Allergens (Allergenic source) | GenBank Accession No. | Aca f 1 Similarity |
|---|---|---|
| Sal k 5 (*Salsola kali*) | ADK22842.1 | 90% |
| Che a 1 (*Chenopodium album*) | G8LGR0.1 | 89% |
| Cro s 1 (*Crocus sativus*) | AAX93750.1 | 88% |
| Ole e 1 (*Olea europaea*) | P19963.2 | 63% |
| Fra e 1 (*Fraxinus excelsior*) | AAQ83588.1 | 62% |
| Syr v 1 (*Syringa vulgaris*) | S43243 | 63% |
| Lig v 1 (*Ligustrum vulgare*) | O82015.2 | 62% |

## Cross-Reaction of Aca f 2 with Other Members of Profilin Protein Family

The findings of amino acid sequence similarity analysis indicated that Aca f 2 has a high degree of similarity with some profilins from the most common allergenic local plants, predominantly Pro j 2 (*P. juliflora* profilin) (96%) [51] (Table 4, Figure 5).

```
                    10         20         30         40         50         60         70         80
            ....|....| ....|....| ....|....| ....|....| ....|....| ....|....| ....|....| ....|....|
Aca f 2   1 MCWQTYVDDH LMCEIEGTNN HLSGAAILGV DGSVWAQSAS FPQFKPDEIA AVVKDFDGPG TLAPTGLHLG GTKYMVIQGE
Pro j 2   1 .S........ .......... ...A...... .......... .........S .......... .......... ..........
Sal k 4   1 .S........ .......... ..TA...... .........N .........S ....E..EA. .......... ..........
Ama r 2   1 .S..A..... ........T. ..T......L .........D .......... .I.E...E.. .......... ..........
Q.suber   1 .S........ ...D...--Q ..TA...I.H .........T ......E.V. .II....E.. S......... ..........
Ole e 2   1 .S........ ...D...--Q ..TA..VI.H .........T ......E.V. .II....E.. S......... ..........

                    90        100        110        120        130
            ....|....| ....|....| ....|....| ....|....| ....|....| ...  aa
Aca f 2  81 PGRFIRGKKG PGGICVKKTG QALVFGIYDE PVTPGQCNMI VERLGDYLVE QGM 133
Pro j 2  81 ..QV...... .......... ...II..... .......... .......... ... 133
Sal k 4  81 A.QV...... .......... ...I...... .......... .......... ... 133
Ama r 2  81 ..AV...... A......... ....M..... .......... ........I. ..Y 133
Q.suber  79 ..AV...... A...T..... ...I...... .L......I. ........I. ..L 131
Ole e 2  79 ..AV...... A...T..... ...I...... .L......I. ........L. ..L 131
```

Figure 5. Comparison of the Aca f 2 (*A. farnesiana* profilin) amino acid sequence with the common allergenic profilins from other plants. The amino acid sequence similarity of Aca f 2 (KM266374) with other members of allergenic profilin family are indicated in Table 4. *Prosopis juliflora* profilin (Pro j 2, AHY24177.1); *Salsola kali* profilin (Sal k 4, ACS34771.1); *Amaranthus retroflexus* profilin (Ama r 2, ACP43298.1); *Quercus suber* profilin (AFG16923.1) and *Olea europaea* profilin (Ole e 2, A4GFC2.1).

This was predictable because *Acacia* and *Prosopis juliflora* both belong to the Fabaceae family. Generally, this high degree of similarity may assume the same tertiary structure [49-51]. Investigation of the amino acid sequences of *A. farnesiana* profilin and several profilin molecules from other plants also revealed cross-reactivity among plant derived profilins from unrelated families, which may be predicted by the degree of amino acid sequence similarity of potential conformational IgE-binding sites (epitopes).

**Table 4. Percentage of similarity between *Acacia* pollen profilin (Aca f 2) and some common allergenic plants profilin**

| Allergens (Allergenic source) | GenBank Accession No. | Aca f 2 Similarity |
|---|---|---|
| Pro j 2 (*Prosopis juliflora*) | AHY24177.1 | 96% |
| Sal k 4 (*Salsola kali*) | ACS34771.1 | 95% |
| Ama r 2 (*Amaranthus retroflexus*) | ACP43298.1 | 93% |
| Q. suber profilin (*Quercus suber*) | AFG16923.1 | 91% |
| Ole e 2 (*Olea europaea*) | A4GFC2.1 | 91% |

# REFERENCES

[1] De Weerd NA, Bhalla PL, Singh MB. Aeroallergens and pollinosis: Molecular and immunological characteristics of cloned pollen allergens. *Aerobiologia*, 2002; 18(2):87-106.

[2] Bush RK. Aerobiology of pollen and fungal allergens. *Journal of Allergy and Clinical Immunology*, 1989; 84(6):1120-24.

[3] D'Amato G, Cecchi L, Bonini S, Nunes C, Annesi-Maesano I, Behrendt H, et al. Allergenic pollen and pollen allergy in Europe. *Allergy*, 2007; 62(9):976-90.

[4] Assarehzadegan MA, Shakurnia A, Amini A. The most common aeroallergens in a tropical region in Southwestern Iran. *World Allergy Organ J*, 2013; 6(1):7.

[5] Bartra J, Mullol J, del Cuvillo A, Davila I, Ferrer M, Jauregui I, et al. Air pollution and allergens. *J Investig Allergol Clin Immunol*, 2007; 17 Suppl 2:3-8.

[6] Beggs PJ. Impacts of climate change on aeroallergens: past and future. *Clin Exp Allergy*, 2004; 34(10):1507-13.

[7] Fereidouni M, Hossini RF, Azad FJ, Assarehzadegan M-A, Varasteh A. Skin prick test reactivity to common aeroallergens among allergic rhinitis patients in Iran. *Allergologia et Immunopatholgia (Madr)*, 2009; 37(2):73-9.

[8] Clot B. Trends in airborne pollen: An overview of 21 years of data in Neuchâtel (Switzerland). *Aerobiologia*, 2003; 19(3-4):227-34.

[9] Damialis A, Halley J, Gioulekas D, Vokou D. Long-term trends in atmospheric pollen levels in the city of Thessaloniki, Greece. *Atmospheric Environment*, 2007; 41(33):7011-21.

[10] Singh AB, Mathur C. An aerobiological perspective in allergy and asthma. *Asia Pacific Allergy*, 2012; 2(3):210.

[11] Keramat A, Marivani B, Samsami M. Climatic Change, Drought and Dust Crisis in Iran. *WASET*, 2011; (57):10-13.

[12] Miri A, Ahmadi H, Ghanbari A, Moghaddamnia A. Dust Storms Impacts on Air Pollution and Public Health under Hot and Dry Climate. *IJEE*, 2007; 1(2):101-05.

[13] Waness A, ABU El-Sameed Y, Mahboub B, Noshi M, Al-Jahdali H, Vats M, et al. Respiratory disorders in the Middle East: A review. *Respiratory*, 2011; 16(5):755-66.

[14] Cakmak S, Dales RE, Coates F. Does air pollution increase the effect of aeroallergens on hospitalization for asthma? *J Allergy Clin Immunol*, 2012; 129(1):228-31.

[15] Lei YC, Chan CC, Wang PY, Lee CT, Cheng TJ. Effects of Asian dust event particles on inflammation markers in peripheral blood and bronchoalveolar lavage in pulmonary hypertensive rats. *Environ Res*, 2004; 95(1):71-6.

[16] Mari A, Schneider P, Wally V, Breitenbach M, Simon-Nobbe B. Sensitization to fungi: epidemiology, comparative skin tests, and IgE reactivity of fungal extracts. *Clin Exp Allergy*, 2003; 33(10):1429-38.

[17] Watanabe M, Igishi T, Burioka N, Yamasaki A, Kurai J, Takeuchi H, et al. Pollen augments the influence of desert dust on symptoms of adult asthma patients. *Allergol Int*, 2011; 60(4):517-24.

[18] Caccavari M, Dome E. An account of morphological and structural characterization of American Mimosoideae pollen. Part I: Tribe Acacieae. *Palynology*, 2000; 24(1):231.

[19] Shamsbiranvand MH, Khodadadi A, Assarehzadegan M-A, Borci SH, Amini A. Immunochemical characterization of *Acacia* pollen allergens and evaluation of cross-reactivity pattern with the common allergenic pollens. *Journal of Allergy (Cairo)*, 2014; 2014:409056.

[20] Ariano R, Panzani RC, Amedeo J. Pollen allergy to mimosa (*Acacia floribunda*) in a Mediterranean area: an occupational disease. *Ann Allergy*, 1991; 66(3):253-6.

[21] Assarehzadegan M-A, Shakurnia AH, Amini A. Sensitization to common aeroallergens among asthmatic patients in a tropical region affected by dust storm. *Journal of Medical Sciences*, 2013; 13(7):592–97.

[22] Bener A, Safa W, Abdulhalik S, Lestringant GG. An analysis of skin prick test reactions in asthmatics in a hot climate and desert environment. *Allergie et Immunologie (Paris)*, 2002; 34(8):281-6.

[23] Suliaman FA, Holmes WF, Kwick S, Khouri F, Ratard R. Pattern of immediate type hypersensitivity reactions in the Eastern Province, Saudi Arabia. *Annals of Allergy Asthma and Immunology*, 1997; 78(4):415-8.

[24] De Zotti R, Gubian F. Asthma and rhinitis in wooding workers. *Allergy Asthma Proc*, 1996; 17(4):199-203.

[25] Wood-Baker R, Markos J. Occupational asthma due to Blackwood (*Acacia* Melanoxylori). *Internal Medicine Journal*, 1997; 27(4):452-53.

[26] Kilpio K, Kallas T, Hupli K, Malanin K. Allergic rhinitis, asthma and eczema caused by gum Arabic in a candy factory worker. *Duodecim; laaketieteellinen aikakauskirja*, 2000; 116(22):2507.

[27] Correia, Osvaldo, M. Antónia Barros, José Mesqutta-Guimarães Airborne contact dermatitis from the woods *Acacia* melanoxylon and Entandophragma cylindricum. *Contact dermatitis*, 1992; 27(5):343-44.

[28] Kung S-J, Mazhani L, Steenhoff AP. Allergy in Botswana: original research article. *Current Allergy & Clinical Immunology*, 2013; 26(4):202-09.

[29] Bousquet J, Cour P, Guerin B, Michel FB. Allergy in the Mediterranean area I. Pollen counts and pollinosis of Montpellier. *Clinical & Experimental Allergy*, 1984; 14(3):249-58.

[30] Baratawidjaja IR, Baratawidjaja PP, Darwis A, Soo-Hwee L, Fook-Tim C, Bee-Wah L, et al. Prevalence of allergic sensitization to regional inhalants among allergic patients in Jakarta, Indonesia. *Asian Pac J Allergy Immunol*, 1999; 17(1):9-12.

[31] Sam CK, Kesavan P, Liam CK, Soon SC, Lim AL, Ong EK. A study of pollen prevalence in relation to pollen allergy in Malaysian asthmatics. *Asian Pac J Allergy Immunol*, 1998; 16(1):1-4.

[32] Liam CK, Loo KL, Wong CM, Lim KH, Lee TC. Skin prick test reactivity to common aeroallergens in asthmatic patients with and without rhinitis. *Respirology*, 2002; 7(4):345-50.

[33] Valenta R, Duchene M, Ebner C, Valent P, Sillaber C, Deviller P, et al. Profilins constitute a novel family of functional plant pan-allergens. *J Exp Med*, 1992; 175(2):377-85.

[34] Mandal J, Roy I, Chatterjee S, Gupta-Bhattacharya S. Aerobiological investigation and in vitro studies of pollen grains from 2 dominant avenue trees in Kolkata, India. *J Investig Allergol Clin Immunol*, 2008; 18(1):22-30.

[35] Valenta R, Kraft D. From allergen structure to new forms of allergen-specific immunotherapy. *Curr Opin Immunol*, 2002; 14(6):718-27.

[36] Valenta R, Kraft D. From allergen structure to new forms of allergen-specific immunotherapy. *Current Opinion in Immunology*, 2002; 14(6): 718–27.

[37] Dhyani A, Singh BP, Arora N, Jain VK, Sridhara S. A clinically relevant major cross-reactive allergen from mesquite tree pollen. *Eur J Clin Invest*, 2008; 38(10):774-81.

[38] Dhyani A, Arora N, Gaur SN, Jain VK, Sridhara S, Singh BP. Analysis of IgE binding proteins of mesquite (*Prosopis juliflora*) pollen and

cross-reactivity with predominant tree pollens. *Immunobiology*, 2006; 211(9):733-40.

[39] Assarehzadegan M-A, Khodadadi A, Amini A, Shakurnia A-H, Marashi SS, Ali-Sadeghi H, et al. Immunochemical characterization of *Prosopis juliflora* pollen allergens and evaluation of cross-reactivity pattern with the most allergenic pollens in tropical areas. *Iranian Journal of Allergy, Asthma and Immunology*, 2014; 14(1):74-82.

[40] Khosravi GR, Assarehzadegan M-A, Morakabati P, Akbari B, Dousti F. Cloning and expression of Aca f 1: a new allergen of *Acacia farnesiana* pollen. *Central-European journal of immunology*, 2016; 41(3):273.

[41] de Dios Alche J, Mrani-Alaoui M, Castro AJ, Rodriguez-Garcia MI. Ole e 1, the major allergen from olive (*Olea europaea* L.) pollen, increases its expression and is released to the culture medium during in vitro germination. *Plant and cell physiology*, 2004; 45(9):1149-57.

[42] Barderas R, Purohit A, Rodriguez R, Pauli G, Villalba M. Isolation of the main allergen Fra e 1 from ash (*Fraxinus excelsior*) pollen: comparison of the natural and recombinant forms. *Annals of Allergy Asthma and Immunology*, 2006; 96(4):557-63.

[43] Batanero E, Gonzalez De La Pena MA, Villalba M, Monsalve RI, Martin-Esteban M, Rodriguez R. Isolation, cDNA cloning and expression of Lig v 1, the major allergen from privet pollen. *Clinical and Experimental Allergy*, 1996; 26(12):1401-10.

[44] Gonzalez E, Villalba M, Rodriguez R. Immunological and molecular characterization of the major allergens from lilac and privet pollens overproduced in Pichia pastoris. *Clinical and Experimental Allergy*, 2001; 31(2):313-21.

[45] Castro L, Mas S, Barderas R, Colas C, Garcia-Selles J, Barber D, et al. Sal k 5, a member of the widespread Ole e 1-like protein family, is a new allergen of Russian thistle (*Salsola kali*) pollen. *International Archives of Allergy and Immunology*, 2014; 163(2):142-53.

[46] Barderas R, Villalba M, Lombardero M, Rodriguez R. Identification and characterization of Che a 1 allergen from Chenopodium album

pollen. *International Archives of Allergy and Immunology*, 2002; 127(1):47-54.

[47] Asturias JA, Arilla MC, Gomez-Bayon N, Martinez J, Martinez A, Palacios R. Cloning and expression of the panallergen profilin and the major allergen (Ole e 1) from olive tree pollen. *Journal of Allergy and Clinical Immunology*, 1997; 100(3):365-72.

[48] Calabozo Bn, Díaz-Perales A, Salcedo G, Barber D, Polo F. Cloning and expression of biologically active *Plantago lanceolata* pollen allergen Pla l 1 in the yeast Pichia pastoris. *Biochemical Journal*, 2003; 372:889-96.

[49] Amini A, sankian M, Assarehzadegan MA, Vahedi F, Varasteh A. *Chenopodium album* pollen profilin (Che a 2): homology modeling and evaluation of cross-reactivity with allergenic profilins based on predicted potential IgE epitopes and IgE reactivity analysis. *Mol Biol Rep*, 2011; 38(4):2579-87.

[50] Assarehzadegan MA, Amini A, Sankian M, Tehrani M, Jabbari F, Varasteh A. Sal k 4, a new allergen of Salsola kali, is profilin: a predictive value of conserved conformational regions in cross-reactivity with other plant-derived profilins. *Biosci Biotechnol Biochem*, 2010; 74(7):1441-6.

[51] Sepahi N, Khodadadi A, Assarehzadegan M, Amini A, Zarinhadideh F, Ali-Sadeghi H. Molecular Cloning and Expression of A New Allergen of *Acacia farnesiana* (Aca f 2). *Iran J Allergy Asthma Immunol*, 2015; 14(4):370-78.

[52] Tehrani M, Sankian M, Assarehzadegan MA, Falak R, Noorbakhsh R, Moghadam M, et al. Identification of a new allergen from *Amaranthus retroflexus* pollen, Ama r 2. *Allergol Int*, 2011; 60(3):309-16.

[53] Assarehzadegan M-A, Amini A, Sankian M, Tehrani M, Jabbari F, Varasteh A. Sal k 4, a new allergen of *Salsola kali*, is profilin: a predictive value of conserved conformational regions in cross-reactivity with other plant-derived profilins. *Biosci Biotechnol Biochem*, 2010; 74(7):1441-6.

[54] Martinez A, Asturias JA, Monteseirin J, Moreno V, Garcia-Cubillana A, Hernandez M, et al. The allergenic relevance of profilin (Ole e 2) from *Olea europaea* pollen. *Allergy*, 2002; 57 Suppl 71:17-23.

[55] Wopfner N, Willeroidee M, Hebenstreit D, van Ree R, Aalbers M, Briza P, et al. Molecular and immunological characterization of profilin from mugwort pollen. *Biol Chem*, 2002; 383(11):1779-89.

[56] Howlett BJ, Hill DJ, Knox RB. Cross-reactivity between *Acacia* (wattle) and rye grass pollen allergens. Detection of allergens in *Acacia* (wattle) pollen. *Clin Allergy*, 1982; 12(3):259-68.

[57] Assarehzadegan M, Shakurnia A, Amini A. The most common aeroallergens in a tropical region in Southwestern Iran. *World Allergy Organ J*, 2013; 6:7.

[58] Furuya K. [Pollinosis. 3. The significance of oak (genus Quercus) in pollinosis]. *Arerugi*, 1970; 19(12):918-30.

[59] Ipsen H, Hansen OC. The NH 2-terminal amino acid sequence of the immunochemically partial identical major allergens of alder (*Alnus glutinosa*) Aln g I, birch (*Betula verrucosa*) Bet v I, hornbeam (*Carpinus betulus*) Car b I and oak (*Quercus alba*) Que a I pollens. *Molecular immunology*, 1991; 28(11):1279-88.

[60] Weber RW. Cross-reactivity of pollen allergens. *Curr Allergy Asthma Rep*, 2004; 4(5):401-8.

[61] Weber RW. Cross-reactivity of pollen allergens: recommendations for immunotherapy vaccines. *Curr Opin Allergy Clin Immunol*, 2005; 5(6):563-9.

[62] Niederberger V, Horak F, Vrtala S, Spitzauer S, Krauth MT, Valent P, et al. Vaccination with genetically engineered allergens prevents progression of allergic disease. *Proc Natl Acad Sci U S A*, 2004; 101 Suppl 2:14677-82.

[63] Weber RW. Cross-reactivity of pollen allergens: impact on allergen immunotherapy. *Ann Allergy Asthma Immunol*, 2007; 99(3):203-11; quiz 12-3, 31.

In: *Acacia*  ISBN: 978-1-53614-237-2
Editor: Aide Matheson  © 2018 Nova Science Publishers, Inc.

**Chapter 7**

# THE GEOGRAPHIC MOSAIC THEORY OF COEVOLUTION APPLIED TO THE NEOTROPICAL MUTUALISM *ACACIA-PSEUDOMYRMEX*

## *Sandra Luz Gómez Acevedo*

Unidad de Morfología y Función. Facultad de Estudios Superiores Iztacala, Universidad Nacional Autónoma de México, Tlalnepantla, Estado de México, Mexico

## ABSTRACT

The symbiotic associations between plants and ants are one of the classic examples of mutualism in nature. This kind of relationship achieves a high degree of sophistication in the myrmecophytic plants, which are characterized by offering permanent shelter and food for their resident ants. In return, the ants protect the plants for the herbivory and confer additional benefits such as protection against pathogenic fungi; the removal of lianas; the addition of nutrients (principally nitrogen) through waste decomposition; and the absorption of $CO_2$. The relationship between swollen thorn *Acacias* and their associated ants has been one of the best known symbiosis so far. The Neotropical *Acacia-Pseudomyrmex* mutualism includes 15 species of *Acacias* and a group of 10 species of mutualistic ants whose geographical distribution is similar. This relationship is frequently cited as an example of coevolution, a term that has been used to refer to the reciprocal change of interacting species, where each of them acts as an agent of natural selection with respect to the other, and where the reciprocal selection would result in congruent phylogenies. However, several reports indicate that the relationship between these taxa does not correspond to the original proposal of strict coevolution, since throughout its geographical distribution, a single ant-*Acacia* species can be inhabited by two or more species of mutualistic ants and even non-mutualistic ants, and have referred to this system as an example of diffuse coevolution. Nevertheless, this kind of relationship shows that although at a population level, relations imply a certain specialization towards one or another species of ant, at a global level, when considering the entire geographical distribution, the evolutionary unit of interaction can include more than one pair of species. This generates a whole range of interrelations, pointing out that the coevolutionary process between these taxa is highly dynamic and corresponds to the proposal of the geographical mosaic theory.

## 1. ANT-PLANT INTERACTIONS

Ecological interactions between species are important evolutionary factors and can be one of the main forces that lead to the diversification and specialization of lineages (Chenuil & McKey, 1996; Fiala et al., 1999). They have fundamented the coevolutionary process, to the point that theories focused on the recognition of ecological interactions, as a fundamental factor in evolutionary change, have been developed; as in the case of the Red Queen theory, where intraspecific ecological relationships are considered the main selection factor (Brockhurst et al., 2014). These interactions may even influence the process of character diversification, establishing fluctuations with respect to the way in which relations between two or more organisms occur. Another influence could be the plasticity of the characters or alleles involved within the relationship (Dercole et al., 2006). However, in nature there are relations that cannot be integrated into a clear pattern due to their dynamics and complexity, such is the case of myrmecophily.

Myrmecophily involves relationships between plants and ants, which have been studied both in a punctual manner and from a theoretical point of view. Pringle and collaborators (2011) propose that the plant-ant interaction depends on the metabolic cost that is offered by the plant, and the benefit and commitment offered by the ants when defending the plants. The dynamics of this interaction, at a theoretical level, establishes that the processes of herbivory will be diminished, as long as a linear relationship exists between the investment of resources by the plant and the growth of resident ant colonies (Fonseca, 1993).

A point to detail about these models, is to recognize that although it has been exemplified in several references as a primordial part of the development of angiosperms, myrmecophily occurs in different groups of plants. That is, although in a plural way it is reported as a process that happens mainly between angiosperms and a few species of ants, it also occurs in species of the *Lecanopteris* and *Platycerium* genera, which are part of the Polypodiaceae family. These ferns are related primarily to tree ants of the species *Crematogaster difformis* (Tanaka & Itioka, 2011). Also, the myrmecophily process also occurs in gymnosperms, particularly in coniferous forest taxa where it is possible to recognize ant nests; niches that also allow the development of different myrmecophilous interactions between invertebrates and degraders (Laakso & Setälä, 1998).

However, as mentioned previously, the greatest number of myrmecophilous species occur within angiosperms, where there could be a total of 1,139 taxa, including participants from at least 50 families; of which the most representative are Leguminosae, Melastomataceae and Rubiaceae (Chomicki & Renner, 2015). In a complementary way, there are at least 25 genera of ants linked to this lifestyle (McKey & Davidson, 1993), and they are related to plants with herbaceous, shrub, arboreal and lianescent life-forms; in tropical, temperate and arid environments around the world (Chomicki & Renner, 2015; Jolivet, 1998; Schupp & Fenner, 1990).

It is important to highlight that this type of ant-plant systems, also provide a large number of examples of mutualisms, which include opportunistic and facultative interactions, ranging from protection against herbivores and seed dispersal to obligate interactions, such as ant-plant symbiosis. Particularly, it is important to note that ant-plant symbiosis is found only in the tropics and includes plant taxa that are generally referred to as ant-myrmecophyte (Mayer et al., 2014), a group we will refer to hereinafter.

In the ant-myrmecophyte group, the plants involved have certain particular morphological characteristics related to the association with ants. The morphological characteristic common to all these plants are the domatia, which are nesting sites for the resident ants. Domatia are cavities that can develop in vegetative organs such as petioles, stems, leaf bases, roots, as well as invaginations or foldings of leaves. These sites occasionally, do not only provide nesting sites for ants, but also provide places in which some species can raise homoptera to supplement their diet (Davidson & McKey, 1993; Mayer et al., 2014). Some authors consider that domatia-bearing plants evolved from plants that maintained a facultative interaction with ants. Throughout the evolutionary history of angiosperms, at least 158 independent origins and 43 losses of domatia have been detected (Chomicki & Renner, 2015).

Another of the characteristics present in myrmecophile plants (although not necessarily in all of them), corresponds to alimentary rewards, which come either from extrafloral nectaries (nectar) or from food bodies rich in nutrients; or both. The chemical constitution of these food rewards offered to ants varies from amino acids and sugar solutions produced in extrafloral nectaries, to proteins or food bodies based on glycogen (Hölldobler & Wilson, 1990; Davidson & McKey, 1993). The extrafloral nectaries are presented as a fundamental source of carbohydrates for worker ants and allow to establish a direct reward mechanism. They are usually located either on the petiole or rachis of the leaves. It has been recognized that in order to link them to a reward mechanism, they must lack alkaloids and cyanogenic glycosides, which are defense mechanisms of plants in general, against phytophagous insects (Rehr et al., 1973).

It has been proposed that in comparison with extrafloral nectaries and food bodies, domatia are the most important factor for the development of obligate myrmecophytic relationships, promoting a greater constancy in the interaction and therefore an increase in the efficacy of the protection

(Chomicki & Renner, 2015; Fiala & Maschwitz, 1992a, b; Heil & McKey, 2003).

In exchange for the domatia and the food offered by the hosts, the ants protect the plants from herbivory and confer additional benefits such as protection against pathogenic fungi; the removal of vines; and the addition of $CO_2$ and nutrients, mainly nitrogen, due to the decomposition of waste (Sagers et al., 2000; Fischer et al., 2003). However, some ant-plant systems of Neotropical and Paleotropical rainforests also involve the presence of other insects (Hemiptera: Coccoidea; Hölldobler & Wilson, 1990; Ward, 1991), which frequently inhabit the plant; either within the domatia, in other hollow stems or in constructions made by the ants on stems or leaves (Heckroth et al., 1998).

Some authors (McKey, 1989; Ward, 1991) consider the presence of coccids as an important factor in the evolution of myrmecophily, since they allow a closer relationship between ants and plants. In turn, the ants are privileged by the food resources provided by such insects, which consist especially of an extra supply of sugars contained in the excreta (Way, 1963), although in some occasions these insects are consumed directly as a source of proteins or lipids (Wheeler, 1942). In counterpart, ants provide vital services for these insects, such as protection against parasites, transport to new nesting sites and an improvement to their sanitary conditions by removing their waste (Way, 1963). Even some founding queen ants bear them in their jaws during their nuptial flight (Buschinger et al., 1987; Klein et al., 1992, cited in Heckroth et al., 1998).

## 1.1. Myrmecophily in the *Acacia s.l.* Genus

Worldwide, myrmecophily shows a greater representation in the tropics, which is where the highest percentage of plant-ant interactions is concentrated. However, when comparing the Paleotropical and Neotropical myrmecophilous flora, there are differences in the number of species and geographical distribution. That is to say, the African tropical myrmecophilous flora is less rich in species, although many of them present a very wide distribution. In contrast, in the Neotropics there is a greater number of genera rich in myrmecophilous species but with restricted geographical distributions. These latter species appear to be entirely neoendemic, the product of a rapid and recent speciation, rather than paleoendemic relicts (Davidson & McKey, 1993).

Also, it has been observed that in the genera that present myrmecophilous species in both the Neotropic and the Paleotropic, myrmecophily is present only in one of the regions. The only exception known to date is in the genus *Acacia sensu lato*, which includes this type of plants both in Africa and in America, and in both regions they extend the syndrome toward arid zones (McKey & Davidson, 1993). This is why it is considered an adequate element of study, in order to recognize both common patterns and differential elements.

The taxonomy of the *Acacia s.l.* genus has been in constant change, it was considered a monophyletic genus that included approximately 1600 species, which were subdivided into three subgenera: *Acacia*, *Aculeiferum* and *Phyllodineae* (Lewis et al., 2005). Recently, the division of the genus into five lineages was proposed, equivalent to five genera, among which is the genus *Vachellia*, which includes all the taxa of the subgenus *Acacia* (Miller & Seigler, 2012), which share a common ancestor and therefore correspond to a monophyletic group (Gómez-Acevedo et al., 2010). Within this subgenus are located four African species (*Acacia bussei*, *A. drepanolobium*, *A. luederitzii* and *A. seyal*; Janzen, 1974) and 15 Neotropical taxa (*A. allenii*, *A. cedilloi*, *A. chiapensis*, *A. cookii*, *A. cornigera*, *A. collinsii*, *A. gentlei*, *A. globulifera*, *A. hindsii*, *A. hirtipes*, *A. janzenii*, *A. mayana*, *A. melanoceras*, *A. ruddiae* and *A. spherocephala*; Rico-Arce, 2003) that maintain a very close relationship with ants. However, despite the nomenclatural change (*Vachellia*), these plants are traditionally known as "ant-*Acacias*" or "swollen-thorn *Acacias*", so we will keep that name throughout this chapter. As can be appreciated, the number of species present in both the Paleotropic and the Neotropic is differentiable. From now on we will refer only to the symbiotic relationship present in the Neotropic.

## 1.2. *Acacia-Pseudomyrmex* Neotropical Mutualism

The relationship between swollen-thorn *Acacias* and their associated ants is one of the best known examples in the field of interactions, and was described in detail by Janzen (1966). This is one of the most constant and persistent mutualisms between plants and insects, which has generated great morphological specialization in both groups. Some studies provide a temporary framework for the origin of this mutualism, placing it in Mesoamerica at the end of the Miocene (Chomicki & Renner, 2015; Gómez-Acevedo et al., 2010; Ward & Branstetter, 2017), with an eventual diversification of both groups towards the north of Mexico; mainly along the coasts towards the Tropic of Cancer (Janzen, 1966). The greatest diversity of the species from both groups involved is found in Mexico, in particular for myrmecophytic *Acacias*. This distribution could have been assisted by birds, who feed on the seeds of these plants (Janzen, 1966, 1969).

Neotropical ant-*Acacias* form a monophyletic group (Gómez-Acevedo et al., 2010), that is distributed collectively in coastal regions from central Mexico to northwestern Colombia, generally at elevations below 1,200 masl. However, only three species have a wide distribution: *A. collinsii* (Mexico to Colombia), *A. cornigera* (Mexico to Costa Rica) and *A. hindsii* (Mexico to Nicaragua). The morphological and chemical characteristics related to the mutualist association with ants include: 1) the presence of domatia, which in this case correspond to modified spiny stipules, which are inhabited by the resident ants; 2) Beltian bodies located at the apex of leaflets; 3) production of new leaves throughout the year to guarantee the presence of Beltian bodies; 4) one or more extrafloral nectaries on the petiole or rachis of the leaves; and 5) lack of chemical compounds such as alkaloids and cyanogenic glycosides, which are defense mechanisms of plants in general, against phytophagous insects (Janzen, 1966; Rehr et al., 1973). The Beltian bodies are constituted by lipids (1 - 10% of dry weight), proteins and free amino acids (8 - 14% of dry weight), carbohydrates (3 - 11% of dry weight) and water (18 - 24%), and they are used as a source of exclusive food for the larvae of the mutualistic ants (Heil et al., 2004).

The populations of ant-*Acacias* show a clinal variation in characteristics such as the size of the leaflets; the size and shape of the spine; life-forms; and the shape of the seed and the pod. Most of the species have increased their local density, population size and geographical coverage; invading farmlands, pastures and roadsides that connect to the disturbed natural sites that are found along rivers, streams and lakes (Janzen, 1974).

In spite of the clinical variations that are present due their distribution, it is possible to recognize basal elements in this plant-ant relationship. An example of this is *Acacia chiapensis*, a species that is located at the base of the myrmecophile group (Gómez-Acevedo et al., 2010), which presents intermediate characteristics in relation to the other myrmecophile *Acacias*, and although it contains chemical compounds that avoid symbiosis, it is capable of surviving in the absence of ants, although in its natural habitat it may be associated with them (Rehr et al., 1973).

It is a fact that the main elements that regulate the coevolution process have as main component the variation of all the characters that ensure this same process, as just mentioned for Neotropical ant-*Acacias*, with their variants and components. Next in this chapter, the interaction between plants and ants in this mutualism is contextualized a bit more.

## 1.3. Mutualist Ants: *Pseudomyrmex ferrugineus* Group

The ants that maintain an obligate mutualism with Neotropical ant-*Acacias* belong to the Neotropical group *Pseudomyrmex ferrugineus*, which consists of 10 species (*P. ferrugineus*, *P. flavicornis*, *P. janzenii*, *P. mixtecus*, *P. nigrocinctus*, *P. particeps*, *P. peperi*, *P. satanicus*, *P. spinicola* and *P. veneficus*); for which it is considered that they are not related to any other group of plants and that they nest exclusively in the domatia of ant-*Acacias* of the New World (Ward, 2017). Janzen (1974) proposed that these ants are not specific to a species of ant-*Acacia*, but to their way of life. This group of ants inhabit from the east and west of Mexico to Central America and northern Colombia; and although there is no species that covers this entire range, its group distribution is similar to that of its hosts; however, only *P. ferrugineus* and *P. peperi* present very wide distributions, from Mexico to Honduras, and to Nicaragua in the second case; while the others occupy more limited localities. These species exhibit a very aggressive behavior and vigorously defend their hosts (Ward, 1993). The most representative species of the group are *P. ferrugineus*, *P. flavicornis*, *P. peperi* and *P. veneficus*, which are mentioned below.

*Pseudomyrmex ferrugineus* is the most common and widely distributed species within the group. Its distribution extends from eastern and southern Mexico to El Salvador and Honduras, and it has been found forming monogynous colonies in the domatia of the 10 Neotropical ant-*Acacias* (*A. chiapensis*, *A. collinsii*, *A. cookii*, *A. cornigera*, *A. gentlei*, *A. globulifera*, *A. hindsii*, *A. janzenii*, *A. mayana* and *A. sphaerocephala*) that are found in this geographical range. *P. flavicornis* is a species similar to *P. ferrugineus*. Their colonies are also monogynous, but their distribution is restricted, from Guatemala to Costa Rica, inhabiting mainly *A. collinsii* and less frequently *A. cornigera* and *A. hindsii* (Gómez-Acevedo et al., 2010; Ward, 1993).

*Pseudomyrmex peperi* is the second most widely distributed species, ranging from eastern Mexico to Nicaragua and has been collected in the domatia of *A. cedilloi*, *A. chiapensis*, *A. collinsii*, *A. cornigera*, *A. globulifera* and *A. gentlei*. It occurs sympatrically with *P. ferrugineus*, but unlike it, forms highly polygynous colonies (Ward, 1993).

*P. veneficus* is characterized by the formation of highly polygynous unicolonies (which include millions of workers and a few hundred queens), which are among the largest of all social insects. This species was considered a taxa of restricted distribution in Mexico, from Sinaloa to Michoacán, inhabiting only *A. collinsii* and *A. hindsii* (Ward, 1993), however, recent reports extend the geographic and host distribution for this ant. It has also been collected in *A. hindsii* in the states of Guerrero, Nayarit and Oaxaca; and in *A. cornigera* in Guerrero in Mexico (Gómez, 2010).

It is important to point out that one of the most fascinating adaptations present in the *Acacia-Pseudomyrmex* Neotropical mutualist system, is found both in the special composition of the nectar and in the digestive system of the associated ants. That is, in the nectaries of the *Acacias* there is simultaneous secretion of sucrose and invertase, an enzyme responsible for hydrolyzing this sugar, in such a way that finally the nectar is constituted mainly of glucose and fructose, which is favorable for the mutualistic ants, who unlike the rest of the ants, lack invertase in their digestive system. However, the loss of invertase is only present in the workers and not in the larvae, in whom the activity of this enzyme has been detected, since they are fed exclusively with Beltian bodies, which contain sucrose (Heil et al., 2004). This adaptation allows in turn to avoid unwanted visitors (Heil et al., 2005). It is necessary to point out that extrafloral nectaries in non-myrmecophytic *Acacias* such as *A. cochliacantha*, *A. farnesiana* and *A. macracantha*, as well as in *Leucaena leucocephala* (Leguminosae), present small amounts of invertase, so that the high amount of this enzyme in the extrafloral nectaries of myrmecophytic *Acacias* represent a quantitative change instead of a qualitative one (Heil et al., 2005; Kautz et al., 2009).

For the *Acacia-Pseudomyrmex* system it has traditionally been reported that there is only one colony of mutualistic ants per *Acacia* individual, however, in *in situ* observations in recent references this behavior has been questioned, since a single *Acacia* individual may be inhabited by two or more species of ants that may be mutualistic and even non-mutualistic (Gómez-Acevedo et al., 2015; Martínez, 2017; Salazar, 2017). The workers of young colonies (less than 50 individuals) leave the domatia only long enough to collect nectar and Beltian bodies, and when the colony consists of 50 to 100 individuals, they begin to patrol all the leaves of their host. Once they have 200 to 400 individuals, which takes about 10 months, the workers become more aggressive and even attack workers from neighboring colonies as well as other insects that reach their host (Janzen, 1967).

## 1.4. *Acacia-Pseudomyrmex* Geographic Mosaic

The relationship between ant-*Acacias* and their associated ants is frequently cited as an example of coevolution, a term that has been used to refer to the reciprocal change of interacting species, where each species acts as a natural selection agent with respect to the other (Thompson, 2005). Traditionally, coevolution has been considered to involve one-to-one interactions in the strict sense, and in these cases the reciprocal selection results in congruent phylogenies (Weiblen & Bush, 2002; Johnson & Clayton, 2004) as in the host-parasite systems *Yuca-Tegeticula* and *Ficus-Agaonide*. However, this type of relationships is the ideal endpoint and could represent the limit of all possible ways in which the coevolutionary process can occur (Thompson, 2005).

However, in nature it is more common to find examples where the evolutionary unit of interaction includes more than one pair of species, that is, the species interact in different ways in different populations, generating a whole range of characteristics that are the result of natural selection. In such cases, host changes are more likely because the taxa involved commonly differ with the species with which they interact. For this type of relations, the terms diffuse coevolution, guild coevolution or multispecific coevolution have been used; concepts that were originally applied in order to consider this type of interactions as not very specific (Thompson, 2005).

The comparison between the phylogenies of ant-*Acacias* and their mutualistic ants of the *P. ferrugineus* group in a recent study (Gómez-Acevedo et al., 2010) revealed the absence of phylogenetic congruency, which is a reflection of the constant change of host by the ants. This depends on the geographical distribution of both groups involved (Figure 1). This scenario is similar to what has been found for other ant-plant mutualistic systems, such as *Azteca-Cecropia* (Ayala et al., 1996); *Crematogaster-Macaranga* (Feldhaar et al., 2003; Quek et al., 2004) and *Petalomyrmex-Leonardoxa* (Chenuil & McKey, 1996).

Ward (1993) proposed that the Neotropical *Acacia-Pseudomyrmex* mutualism represents a good example of diffuse coevolution, because the *Acacias* may be associated at the same time and space with more than one ant species. Likewise, host ants can also inhabit different species of myrmecophile *Acacias* in time and space. In general terms, when applying the term of diffuse coevolution, it tends to be considered that the relations between the participants are lax and that therefore they would not be exerting a strong selection pressure towards their counterpart. Reason why Thompson (2003) proposes the "Theory of the Geographic Mosaic", an alternative that explains in a more precise way what it means to interact and evolve simultaneously with more than one species.

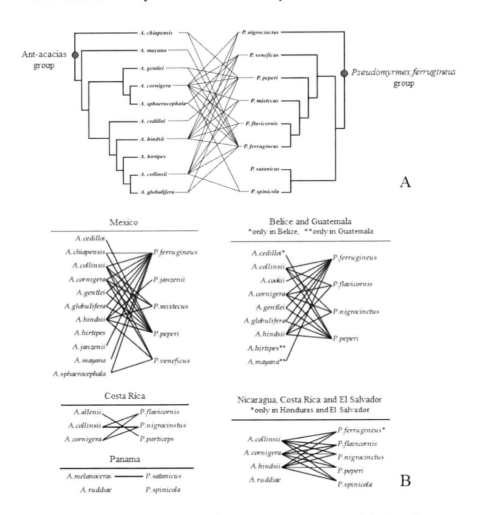

Figure 1. Comparison of the relationships between ant-*Acacias* and the *Pseudomyrmex ferrugineus* group (modified from Gómez-Acevedo et al., 2010). A) Phylogenetic level. B) Geographical level. The lines indicate the relationships reported so far in the literature. For *Acacia hirtipes* and *A. ruddiae* there is no information on associated mutualist ants. *Pseudomyrmex satanicus* is considered the only inhabitant of *Acacia melanoceras*.

In this theory, it is basically proposed that the coevolutionary process is highly dynamic, so that adaptations may appear or disappear because some populations become highly specialized with respect to an interactor, while in other populations such interaction may be less specialized. Also, some of these populations could be outside the geographical distribution area of their interacting species, which would cause either the probable loss of some of the adaptations that allow them to interact or the emergence of new adaptations to establish relationships with other species. Therefore, the global course of the coevolution between two or more species is driven by this geographic ever-changing mosaic of interactions. The geographical point of view is necessary for the understanding of coevolutionary dynamics, because comparisons between local interactions with variant results can help extract the factors that promote or prevent the escalation of coevolutionary characteristics (Benkman et al., 2003; Siepielski & Benkman 2004; Thompson 2005; cited in Toju & Sota, 2006).

The Neotropical mutualist system (*Acacia-Pseudomyrmex*) offers a good opportunity to apply this theory. For example, the ants with greater geographic distributions, such as *P. ferrugineus* and *P. peperi*, nest in a greater number of ant-*Acacias*. In contrast, species with a smaller distribution such as *P. particeps* (restricted to Costa Rica), *P. satanicus* (restricted to Panama) and *P. janzenii* (restricted to western Mexico), have been observed only in one *Acacia* species. These may probably represent examples of local coevolution on the part of these interactors (Ward, 1993). The scenario can be even more complicated, because these plants can be inhabited simultaneously by both mutualist and non-mutualist ants; and even by ants of other genera, as in the myrmecophytic *A. cedilloi*, where in addition to the mutualist *P. peperi*, three more species have been found (*Cephalotes* sp., *Crematogaster torosa* and *P. gracilis*). Likewise, it also draws attention the coexistence of *P. mixtecus* and *P. subtilissimus* with ants of the genus *Crematogaster*. This had not been previously reported and could imply a new association for Neotropical ant-*Acacias* (Gómez, 2010).

It should be noted that *Acacia cedilloi* is a little known Neotropical myrmecophytic whose distribution is restricted to a few fragmented populations in the Yucatán Peninsula (Rico-Arce, 1994) and Belize (L. Rico, personal communication). This species is particularly interesting since it represents the only Neotropical ant-*Acacia*, so far, in which there has been found a greater number of non-mutualistic ants belonging to different *Pseudomyrmex* genera. Figueroa-Castro and Castaño-Meneses (2004) conducted a study to evaluate the myrmecofauna associated with this species. They analyzed 129 plants, of which only 24.81% lacked ants. In the other individuals, they found in total 161 ants belonging to three subfamilies, four genera and six species: *Cephalotes minutus*, *Crematogaster curvispinosa*, *C. torosa*, *Ectatomma tuberculatum*, *Pseudomyrmex ferrugineus* (which was the most abundant) and *P. gracilis*. These authors mention that *Ectatomma tuberculatum* occupies *A. cedilloi* to forage extrafloral nectar and occasionally to hunt *Pseudomyrmex* individuals, which they carry in their jaws back to their nests.

Likewise, Raine et al. (2004) conducted a detailed study in an *Acacia mayana* population, where they found that 76.5% of 34 individuals studied were exclusively occupied by *P. ferrugineus* (mutualistic species), while in the remaining 23.5% this same species was found coexisting with *Camponotus planatus*, which represents a mutualism parasite, since it does not defend its host. It should be noted that in the individuals that had both species of ants, the mutualist *P. ferrugineus* occupied mainly the domatia near the new branches, which is where the Beltian bodies are basically found, because these structures are used exclusively to feed the pupae and once they are cut from the leaves, they are not produced again by the plants.

In a similar study, Clement et al. (2008) studied four species of Neotropical myrmecophile *Acacias* (*A. chiapensis*, *A. collinsii*, *A. cornigera* and *A. hindsii*), in which they found as a resident ant the mutualist *P. ferrugineus*, coexisting with other species of the genus *Pseudomyrmex*, as well as with ants of other genera. Regarding their spatial distribution, they observed that plants less than 1 meter high, could be occupied by more than one species of ant, although in separate branches. The most extreme case was *A. hindsii*, where they observed three species of mutualistic ants (*P. ferrugineus*, *P. mixtecus* and *P. peperi*) and two non-mutualists (*P. gracilis* and *P. nigropilosus*). As in the work reported by Raine et al. (2004), non-mutualistic ants were found inhabiting the lower parts of the plants, which are not visited by resident mutualist ants.

## Conclusion

Previously, the importance of the *Acacia s.l.* genus in the study of myrmecophily in both tropical regions (Paleotropic and Neotropic) had been commented on. However, in recent reviews a particular link has been recognized that allows us to identify fundamental elements, which lead to the integration and recognition of the Neotropical *Acacia-Pseudomyrmex* system as a fluctuating and dynamic mutualism.

This mosaic of interactions can be appreciated when comparing the phylogenies of both groups. With this comparison it is possible to appreciate not only in a general way the relationships and characteristics that have been fixed over time in all populations of these species; but it also reveals that both coevolution (observed in the physiological modifications of the ants; the morphological adaptations in the plants; and the estimated dates of divergence) and intrageneric colonization have been two processes of great importance that have taken place during the evolution of this mutualism.

Finally, with the comparison of phylogenies it is possible to appreciate all the possible relationships reported so far, at least for the groups that make up the obligate Neotropical mutualism. It is possible to verify that the coevolutionary process is much more dynamic than what is expressed in studies of distribution or of individual populations. The observable structuring demonstrates, as Thompson pointed out in his Geographic Mosaic theory, that the general evolution of a species is the product of coevolution with several other species, although the individual populations are specialized only towards one or two taxa.

## Acknowledgments

This work was supported by UNAM-PAPIIT IN217217.

## REFERENCES

Ayala, F. J., Wetterer, J. K., Longino, J. T. & Hartl, D. L. (1996). Molecular phylogeny of *Azteca* ants (Hymenoptera: Formicidae) and the colonization of *Cecropia* trees. *Molecular Phylogenetics and Evolution*, 5, 423-428.

Brockhurst, M. A., Chapman, T., King, K. C., Mank, J. E., Paterson, S. & Hurst, G. D. D. (2014). Running with the Red Queen: the role of biotic conflicts in evolution. *Proceedings of the Royal Society B*, 281(1797), 20141382. https://doi.org/10.1098/rspb.2014.1382.

Chenuil, A. & McKey, D. (1996). Molecular phylogenetic study of a myrmecophyte symbiosis: did *Leonardoxa*/ant associations diversify via cospeciation? *Molecular Phylogenetics and Evolution*, 6, 270–286.

Chomicki, G. & Renner, S. S. (2015). Phylogenetics and molecular clocks reveal the repeated evolution of ant-plants after the late Miocene in Africa and the early Miocene in Australasia and the Neotropics. *New Phytologist*, 207(2), 411–424. https://doi.org/10.1111/nph.13271.

Clement, L. W., Köppen, S. C., Brand, W. A. & Heil, M. (2008). Strategies of a parasite of the ant-*Acacia* mutualism. *Behavioral Ecology and Sociobiology*, 26, 953-962.

Davidson, D. W. & McKey, D. (1993). The evolutionary ecology of symbiotic ant-plant relationships. *Journal of Hymenopteral Research*, 2 (1), 13-83.

Dercole, F., Ferrière, R., Gragnani, A. & Rinaldi, S. (2006). Coevolution of slow-fast populations: evolutionary sliding, evolutionary pseudo-equilibria and complex Red Queen dynamics. *Proceedings of the Royal Society B*, 273 (1589), 983–990. https://doi.org/10.1098/rspb.2005.3398.

Feldhaar, H., Fiala, B., Gadau, J., Mohamed, M. & Maschwitz, U. (2003). Molecular phylogeny of *Crematogaster* subgenus *Decacrema* ants (Hymenoptera: Formicidae) and the colonization of *Macaranga* (Euphorbiaceae) trees. *Molecular Phylogenetics and Evolution*, 27, 441-452.

Fiala, B., Jacob, A., Maschwitz, U. & Linsenmair, K. E. (1999). Diversity, evolutionary specialization and geographic distribution of a mutualistic ant-plant complex *Macaranga* and *Crematogaster* in South East Asia. *Biological Journal of the Linnean Society*, 66, 305-331.

Fiala, B. & Maschwitz, U. (1992a). Domatia as most important adaptations in the evolution of myrmecophytes in the paleotropical tree genus *Macaranga* (Euphorbiaceae). *Plant Systematics and Evolution*, 180, 53-64.

Fiala, B. & Maschwitz, U. (1992b). Food bodies and their significance for obligate ant-association in the tree genus *Macaranga* (Euphorbiaceae). *Botanical Journal of the Linnean Society*, 110, 61-75.

Figueroa-Castro, D. M. & Castaño-Meneses, G. (2004). Hormigas (Hymenoptera: Formicidae) asociadas a *Acacia cedilloi* L. Rico (Mimosaceae) en la reserva ecológica "El Edén", Quintana Roo, México. *Entomología Mexicana*, 3, 208-211.

Fischer, R. C., Wanek, W., Ritcher, A. & Mayer, V. (2003). Do ants feed plants? A $^{15}$N labeling study of nitrogen fluxes from ants to plants in the mutualism of *Pheidole* and *Piper*. *Journal of Ecology*, 91, 126-134.

Fonseca, C. R. (1993). Nesting Space Limits Colony Size of the Plant-Ant Pseudomyrmex concolor. *Oikos*, 67(3), 473. https://doi.org/10.2307/3545359.

Gómez, S. (2010). *Estudio comparativo de las tasas de diversificación de los subgéneros neotropicales de Acacia (Leguminosae: Mimosoideae) en México*. Universidad Nacional Autónoma de México. [*Comparative study of the diversification rates of neotropical Acacia (Leguminosae: Mimosoideae) subgenera in Mexico*. National Autonomous University of Mexico].

Gómez-Acevedo, S., Rico-Arce, L., Delgado-Salinas, A., Magallón, S. & Eguiarte, L. E. (2010). Neotropical mutualism between *Acacia* and *Pseudomyrmex*: phylogeny and divergence times. *Molecular Phylogenetics and Evolution*, 561, 393–408. doi:10.1016/j.ympev.2010.03.018.

Gómez-Acevedo, S., Trujillo, C. O., Castaño-Meneses, G., Martínez, T. J., Callejas-Chavero, A., Cueva del Castillo, M. R. & Tapia-Pastrana, F.

(2015). Mirmecofauna asociada a *Acacias* mirmecófilas en Veracruz y Oaxaca, México. [Mirmecofauna associated with Acacias mirmecófilas in Veracruz and Oaxaca, Mexico] In G. Castaño-Meneses, M. Vásquez-Bolaños, J. L. Navarrete-Heredia, G. A. Quiroz-Rocha, & I. Alcalá-Martínez (Eds.), *Avances de Formicidae de México*, (1st ed., pp. 171-177), Juriquilla, Querétaro, Universidad Nacional Autónoma de México & CUCBA, Universidad de Guadalajara.

Heckroth, H. P., Fiala, B., Gullan, P. J., Idris, A. H. & Maschwitz, U. (1998). The soft scale (Coccidae) associates of Malaysian ant-plants. *Journal of Tropical Ecology*, *14*, 427- 443.

Heil, M. & McKey, D. (2003). Protective ant-plant interactions as model systems in ecological and evolutionary research. *Annual Review of Ecology Evolution and Systematics*, *34*, 425-453.

Heil, M., Baumann, B., Krügeeer, R. & Linsenmair, E. (2004). Main nutrient compounds in food bodies of Mexican *Acacia* ant-plants. *Chemoecology*, *14*, 45-52.

Heil, M., Rattke, J. & Bland, W. (2005). Post-secretory hydrolysis of nectar sucrose and specialization in ant-plant mutualism. *Science*, *308*, 560-563.

Hölldobler, B. & Wilson, E. O. (1990). *The ants*. Berlin, Heidelberg, Springer Verlag.

Janzen, D. H. (1966). Coevolution of mutualism between ants and *Acacias* in Central America. *Evolution*, *20* (3), 249-275.

Janzen, D. H. (1967). Interaction of the bulls-horn *Acacia* (*Acacia cornigera* L.) with an ant inhabitant (*Pseudomyrmex ferruginea* F. Smith) in eastern Mexico. *Kansas University Science Bulletin*, *47*, 315-558.

Janzen, D. H. (1969). Birds and the Ant x *Acacia* interaction in Central America, with notes on birds and other myrmecophytes. *Condor*, *71*, 240-256.

Janzen, D. H. (1974). Swollen-thorn *Acacias* of Central America. *Smithsonian Contributions to Botany*, *13*, 1-131.

Johnson, K. P. & Clayton, D. H. (2004). Untangling coevolutionary history. *Systematic Biology*, *53*(1), 92-94.

Jolivet, P. (1998). *Interrelationships between insects and plants*. Boca Raton, Florida, CRC Press, USA.

Kautz, S., Lumbsch, H. T., Ward, P. S. & Heil, M. (2009). How to prevent cheating: a digestive specialization ties mutualistic plant-ants to their ant-plant partners. *Evolution*, *63*(4), 839-853.

Laakso, J. & Setälä, H. (1998). Composition and trophic structure of detrital food web in ant nest mounds of *Formica aquilonia* and in the surrounding forest soil. *Oikos*, *81*, 266-278.

Lewis, G. P., Schrire, B., Mackinder, B. & Lock, M. (2005). *Legumes of the World*. Kew, Royal Botanic Gardens.

Martínez, T. J. (2017). *Hormigas asociadas a Acacia cornigera y Acacia hindsii (Leguminosae-Mimosoideae) en Santiago Pinotepa Nacional, Oaxaca*. Universidad Nacional Autónoma de México. [Ants associated with Acacia cornigera and Acacia hindsii (Leguminosae. Mimosoideae) in Santiago Pinotepa Nacional, Oaxaca. National Autonomous University of Mexico].

Mayer, V. E., Frederickson, M. E., McKey, D. & Blatrix, R. (2014). Current issues in the evolutionary ecology of ant-plant symbioses. *New Phytologist*, *202*, 749-764.

McKey, D. (1989). Interactions between plants and leguminous plants. In C. H. Stirton, & J. L. Zarucchi (Eds.), *Advances in Legume Biology. Monographs in Systematic Botany from the Missouri Botanical Garden*, *29*, 673-718.

McKey, D. & Davidson, D. W. (1993). Ant-plant symbioses in Africa and the Neotropics: History, biogeography and diversity. In Goldblatt (Ed.), *Biological relationships between Africa and South America*, (pp. 568-606). New Haven, Connecticut, Yale University Press.

Miller, J. T. & Seigler, D. (2012). Evolutionary and taxonomic relationships of *Acacia s.l.* (Leguminosae: Mimosoideae). *Australian Systematic Botany*, *25*, 217-224.

Pringle, E. G., Dirzo, R. & Gordon, D. M. (2011). Indirect benefits of symbiotic coccoids for an ant-defended myrmecophytic tree. *Ecology*, *92* (1), 37–46. https://doi.org/10.1890/10-0234.1.

Quek, S. P., Davies, S. J., Itino, T. & Pierce, N. E. (2004). Codiversification in an ant-plant mutualism: stem texture and the evolution of host use in *Crematogaster* (Formicidae: Myrmicinae) inhabitants of *Macaranga* (Euphorbiaceae). *Evolution*, *58*(3), 554-570.

Raine, N. E., Gammans, N., MacFadyen, I. J., Scrivner, G. K. & Stone, G. N. (2004). Guards and thieves: antagonistic interactions between two ant species coexisting on the same ant-plant. *Ecological Entomology*, *29*, 345-352.

Rehr, S. S., Feeny, P. P. & Janzen, D. H. (1973). Chemical defense in Central American non-ant *Acacias*. *Journal of Animal Ecology*, *42*, 405-416.

Rico-Arce, L. (2003). Geographical patterns in neotropical *Acacia* (Leguminosae: Mimosoideae). *Australian Systematic Botany*, *16*, 41-48.

Rico-Arce, M. de L. (1994). Nueva especie mirmecófila de *Acacia* (Leguminosae) de la Península de Yucatán. *Acta Botánica Mexicana*, *26*, 7-10. [New myrmecophilic species of *Acacia* (Leguminosae) from the Yucatan Peninsula.]

Sagers, C. L., Singer, S. M. & Evans, R. D. (2000). Carbon and nitrogen isotopes trace nutrient exchange in an ant-plant mutualism. *Oecologia*, *123*, 582-586.

Salazar, R. C. (2017). *Estructura de la población de hormigas asociadas a Acacias mirmecófilas en dos poblaciones de México*. Universidad Nacional Autónoma de México. [*Structure of the populations of ants associated with ant-Acacias in two populations of Mexico*. National Autonomous University of Mexico].

Schupp, E. W. & Feener, D. H. (1990). Phylogeny, life form and habitat dependence of ant-defended plants in a Panamiam forest. In C. R. Huxley, & D. F. Cutler (Eds.), *Ant-plant interactions*, (pp. 250-259). New York, EEUU, Oxford Science Publications.

Tanaka, H. O. & Itioka, T. (2011). Ants inhabiting myrmecophytic ferns regulate the distribution of lianas on emergent trees in a Bornean tropical rainforest. *Biology Letters*, *7*(5), 706–709. https://doi.org/10.1098/rsbl.2011.0242.

Thompson, J. N. (2005). *The Geographic Mosaic of Coevolution*. Chicago, University of Chicago Press.

Toju, H. & Sota, T. (2006). Phylogeography and the geographic cline in the armament of a seed-predatory weevil: effects of historical events vs. natural selection from the host plant. *Molecular Ecology*, *15*, 4161-4173.

Ward, P. S. (1991). Phylogenetic analysis of pseudomyrmecine ants associated with domatia-bearing plants. In C. R. Huxley, & D. F. Cutler (Eds.), *Ant-Plant Interactions*, (pp. 335-352). Oxford, Oxford University Press.

Ward, P. S. (1993). Systematic studies on *Pseudomyrmex Acacia*-ants (Hymenoptera: Formicidae: Pseudoyrmecinae). *Journal of Hymenopteral Research*, *2*, 117-168.

Ward, P. S. & Branstetter, M. G. (2017). The *Acacia* ants revisited: convergent evolution and biogeographic context in an iconic ant/plant mutualism. *Proceedings of the Royal Society B*, *284*, 20162569.

Ward, P. S. (2017). A review of the *Pseudomyrmex ferrugineus* and *Pseudomyrmex goeldii* species groups: *Acacia*-ants and relatives (Hymenoptera: Formicidae). *Zootaxa*, *4227*, 524-542.

Way, M. J. (1963). Mutualism between ants and honeydew-producing Homoptera. *Annual Review of Entomology*, *8*, 307-344.

Weiblen, G. D. & Bush, G. L. (2002). Speciation in fig pollinators and parasites. *Molecular Ecology*, *11*, 1573-1578.

Wheeler, W. M. (1942). Studies of Neotropical ant-plants and their ants. *Bulletin of the Museum of Comparative Zoology*, *90*, 1-262.

## BIOGRAPHICAL SKETCH

### *Sandra Luz Gómez Acevedo*

**Affiliation:** Unidad de Morfología y Función, Facultad de Estudios Superiores Iztacala, Universidad Nacional Autónoma de México. Tlalnepantla, Estado de México, México.

**Education:** PhD in Biological Sciences, Master Degree in Systematics, Bachelor Degree in Biology (all by the National Autonomous University of Mexico)

**Research and Professional Experience:** Evolutionary Biology, Floral development, Interactions ant-plant.

In: *Acacia*
Editor: Aide Matheson
ISBN: 978-1-53614-237-2
© 2018 Nova Science Publishers, Inc.

Chapter 8

# POTENTIAL OF *ACACIA MELANOXYLON* FOR PULPING

*Helena Pereira*[1,*], *Rogério Simões*[2], *António Santos*[1], *Jorge Gominho*[1], *Ana Lourenço*[1] *and Ofélia Anjos*[1,3,4]

[1]Centro de Estudos Florestais, Instituto Superior de Agronomia, Universidade de Lisboa, Lisboa, Portugal
[2]Unidade de Materiais Fibrosos e Tecnologias Ambientais (FibEnTech-UBI), Universidade da Beira Interior, Covilhã, Portugal
[3]Instituto Politécnico de Castelo Branco, Castelo Branco, Portugal
[4]Centro de Biotecnologia de Plantas da Beira Interior, Castelo Branco, Portugal

## ABSTRACT

Most of the fibre raw materials used by the pulp and paper industry are from a small number of tree species. For instance, *Eucalyptus* and *Pinus* species are the major industrial pulpwood sources obtained from forests that are characterized by a relatively low biodiversity. The large monoculture areas also increase environmental risks such as those related

---

[*] Corresponding Author Email: hpereira@isa.ulisboa.pt.

to biotic attacks or forest fires. Diversification of industrial fibre sources has therefore been a matter of research and the characterization of different raw materials has been made in view of their pulping potential.

*Acacia melanoxylon* R. Br. (blackwood) grows well in Portugal in pure or mixed stands with *Pinus pinaster* Aiton, and is valued as a timber species with potential for sawmills. In addition, the wood anatomical and chemical characteristics also allow to consider the species as an alternative raw material for the pulp industry. *Acacia* species, with their relatively short, flexible and collapsible fibres, have potential to produce papers with good trade-offs between light scattering/tensile strength and smoothness/tensile strength, at low energy consumption in refining. The pulping and paper making potential of blackwood has been studied by several authors who showed an overall good pulping aptitude under the same experimental conditions of kraft pulping as used for eucalypt pulping, with pulp yields ranging between 47% and 58%. The presence of heartwood should be taken into account because it decreases the raw-material pulping quality due to the higher extractives content. Heartwood proportion should therefore be considered as a quality variable when using *A. melanoxylon* wood in pulp industries.

This chapter characterizes the *A. melanoxylon* wood pulping performance, regarding yield and kappa number, and the pulp and paper properties. The application of fast spectroscopic technologies for pulp quality determination is also described.

**Keywords:** *Acacia melanoxylon*, pulp yield, Kraft pulp, fibres morphology, papermaking potential

## 1. Introduction

Several hardwoods, including some *Acacia* species, were studied for pulp production since the 80's, and extensive plantations were introduced in many countries such as Brazil, South Africa, Indonesia, Vietnam, Malaysia and Australia (Logan and Balodis 1982; Coleman 1998; Jahan et al. 2008). *Acacia mangium* is the main *Acacia* planted in Indonesia, Vietnam and Malaysia, with a total plantation area of 1,400 thousand ha, while *A. mearnsii,* initially exploited for tannin production, is now used for the pulp and paper industries with plantations mainly in South Africa (120,000 ha) and Brazil (250,000 ha) (Griffin et al. 2011).

Indonesia is the largest producer of *Acacia* pulp, mainly bleached kraft pulp (BAKP) that is used in a wide variety of paper grades, due to its excellent sheet formation, opacity and fibre properties. This short-fibre pulp can also be mixed with eucalypt pulps for the production of fine papers (Paavilainen 2000).

Several studies have addressed the wood pulping aptitude of different *Acacias* using various processes such as kraft, soda-anthraquinone (soda-AQ), modified sulfite-AQ (ASA), and sulfite-AQ-methanol (ASAM). The pulp yields ranged from 42% to 56%, with pulp kappa numbers between 11 and 22, and ISO brightness between 20% and 45%. The studied *Acacia* species included *A. auriculiformis* (Jahan et al. 2008), *A. melanoxylon* (Lourenço et al. 2008, Anjos et al. 2011, Santos et al. 2012a), *A. mangium* (Watanabe and Miyanishi 2004) and *A. mearnsii* (Furtado et al. 2000).

The available studies show that *A. melanoxylon* has good aptitude as feedstock for pulp production under different delignification processes (Anjos et al. 2011; Santos et al. 2006, 2012a; Lourenço et al. 2008; Pereira et al. 2016). *A. melanoxylon* wood has already an established commercial timber value given by its heartwood content and golden-brown colour that is quite appreciated as cabinet timber (Searle 2000) and is used for different solid wood products e.g., furniture and woodcrafts (Griffin et al. 2011), building and carpentry (Nicholas and Brown 2002), panel products (Pedieu et al. 2009), and domestic fuelwood (Searle 2000; Griffin et al. 2011)

The aim of this chapter is to provide a review on the pulpwood quality of *A. melanoxylon* by describing the characteristics of the raw material that are relevant for a pulp targeted application, its aptitude for pulping and paper making potential, and the prediction of paper properties.

## 2. WOOD PULPING QUALITY

The requirements that one lignocellulosic material has to meet in order to be considered a good feedstock for the pulp and paper industries are primarily: to be abundant and with high biomass productivity, to have a low cost of installation and production, and to present a high content in cellulose.

The aptitude for pulping and the quality of the pulps are also strongly dependent of the wood anatomy, fibre morphology and chemical composition (Bektas 1999).

One fundamental requirement for a pulpwood species is that it can be planted in managed commercial plantations. At present this is not possible with *A. melanoxylon* in Portugal since the species is considered an invasive and has therefore planting limitations. However, the need for eradication of this species creates an opportunity for the valorisation of the stem by using it as raw material for pulping.

One important quality trait is the proportion and characteristics of heartwood in a pulpwood feedstock (Miranda et al. 2007). The presence of heartwood in *A. melanoxylon* is high, representing 47% to 61% of the total tree volume of mature trees and is a stem quality parameter that has to be taken into account (Knapic et al. 2006; Igartúa et al. 2017). In fact, for wood sawn products the presence of a high proportion of heartwood is much appreciated and therefore it is an advantage, but the same cannot be said for pulp production because heartwood has a high content of extractives which negatively influence the delignification and bleaching process (Hillis, 1987).

In *A. melanoxylon* the heartwood, compared with sapwood, has higher content of extractives (8.2% vs. 4.1%), less lignin (20.7% vs. 21.5%) but a similar carbohydrate content (Lourenço et al. 2008). The presence of more extractives is negative for pulping: heartwood pulps have higher residual lignin in comparison with sapwood pulps (kappa numbers 11 vs. 7) (Lourenço et al. 2008).

Wood density is an important physical parameter that is key in many aspects of wood processing (e.g., chipping, transport and pulping) and product quality (Santos et al. 2008; Silva et al. 2009, Anjos et al. 2011). The pulp yield and Kappa number of *A. melanoxylon* were not correlated with wood density but their variation at a given wood density allows selection of trees for a more efficient pulp production (Santos et al. 2012a).

Fibre characteristics strongly influence the physical and mechanical properties of paper and paperboard (Malan et al. 1994; El-Hosseiny et al. 1999). Fibre length influences the paper sheet formation and its uniformity, and its correlation with tear and tensile index is well established (Paavilainen 2000; Niskanen 1998). The range of pulp fibre length of *A. melanoxylon* from 0.660 to 0.940 mm, and of fibre width from 16.2 to 22.9 µm (Santos et al. 2012a; Pereira et al. 2016) compares favourably with that of other important pulpwoods such as *E. globulus*.

Concerning the wood chemical composition, the content of lignin, cellulose and extractives are important quality parameters of pulpwoods e.g., a low content of lignin and correspondingly a high cellulose content and a low extractives content are favourable characteristics. *A. melanoxylon* wood has an overall low lignin content (18%-20%) and moderate extractives content (3%-6%) leading to a good pulping ability with high kraft pulp yields and well delignified pulps (Lourenço et al. 2008, Santos et al. 2008).

## 3. PULPING

The potential of a given raw material for pulp and paper production is given in a first assessment by the pulping yield and the required cooking conditions, in conjunction with the wood basic density and tree mean annual increment. These are the first quality parameters to be determined, usually before chemical or fibre biometric features, because they give a first estimate of the potential pulp production per unit of wood raw-material.

Several studies investigated the potential of *A. melanoxylon* in kraft pulping (Anjos et al. 2011; Santos et al. 2006; Santos et al. 2012a). In general, it can be said that *A. melanoxylon* has a very good pulping potential although the comparison between results may be difficult due to the different growing conditions, tree age and cooking conditions. The pulp yield ranged from 47.7% to 57.5%, which is in line with pulp yields obtained for the pulping of high performing hardwood species, such as *Eucalyptus globulus* with pulp yields in the range 52-59% (Valente et al. 1992; Miranda et al. 2007). Sapwood performed better in pulping than heartwood with higher yields of 56% in comparison to 53% of heartwood (Lourenço et al. 2008).

**Table 1. Comparison of pulping performance of wood of different *Acacia* species and, whenever available, of *Eucalyptus globulus,* regarding pulp yield, rejects, kappa number and viscosity and alkali consumption**

|  | Pulp yield (%) | Uc (%) | PKN | PV (cm³/g) | AC % (E.A.) | Pulping condition | Reference |
|---|---|---|---|---|---|---|---|
| *A. melanoxylon*, 22-year-old, stem base, b.d.: 460 kg/m³ | 55.7 | 4.5 | 11.0 | 1089** | 14.6 | EA: 20.1% on wood<br>S = 23%; L/W = 4/1<br>$t_{170°C}$ = 70 min | Gil et al. 1999 |
| *E. globulus*, 8-year-old, clone, b.d.: 460 kg/m³ | 52.5 | 0.1 | 12.2 | 1100** | 15.3 |  |  |
| *A. melanoxylon*, 22-year-old, top, b.d.: 387 kg/m³ | 53.2 | 0.4 | 10.9 | 980** | 15.1 | EA:18.7% on wood<br>S = 30%; L/W = 4/1<br>$t_{160°C}$ = 120 min | Santos et al. 2006 |
| *E. globulus*, industrial chips, b.d.: 536 kg/m³ | 50.5 | 0.3 | 14.1 | 956** | 15.2 |  |  |
| *A. melanoxylon*, 45-year-old | 47.7-57.7 | 0.7-1.7 | 13.2-16.8 | 863-1026 * | 15.8-18.9 | EA: 19.6-21.3% on wood<br>S = 30%; L/W = 4/1<br>$t_{160°C}$ = 80-90 min | Anjos et al. 2011 |
| *A. aulacocarpa*, 12-year-old, b.d.: 598 kg/m³ | 55.4 | 0.6 | 19.3 | n.a. | n.a. | AA: 18% on wood<br>S = 25%; L/W = 3.5/1<br>$t_{170°C}$ = 120 min | Clark et al. 1991 |
| *A. mangium*, 6-year-old, b.d.: 490 kg/m³ | 54.6 | n.a. | 19.7 | n.a. | n.a. | AA: 18% on wood | Pan and Lu 1988 |
| *A. mangium*, 9-year-old, 420 kg/m³ | 52.3 | n.a. | 21.0 | n.a. | n.a | A.A.: 18% on wood | Logan and Balodis 1982 |
| *A. mangium*, 8-year-old, b.d.: 571 kg/m³ | 50.8 | 0.86 | 22.4 | n.a. | 17.5 | AA: 25.2% on wood<br>$t_{170°C}$ = 60 min | Griffin et al. 2014 |

*viscosity of bleached pulp; ** viscosity of pulp; AC - alkali consumption: Uc - uncooked; PKN- pulp kappa number; PV - pulp viscosity; AA and EA active and effective alkali, expressed as NaOH; bd – basic density; S – sulfidity; L/W - ratio liquor wood; n.a.: not available

The pulping aptitude of *Acacia* was also analysed using other processes - soda-anthraquinone (soda-AQ), modified sulfite-AQ (ASA), and sulfite-AQ-methanol (ASAM) and the results showed that the species has potential to produce short fibre pulps with good quality: pulp yield from 42% to 56%, kappa numbers between 11 and 22, ISO brightness between 20% to 45% (Santos et al. 2006; 2012a; Lourenço et al. 2008).

**Table 2. Wood density and pulp fibre characteristics of *Acacia melanoxylon* and *Eucalyptus globulus***

|  | A. melanoxylon | | | | E. globulus | |
|---|---|---|---|---|---|---|
| Years | 40 | | 22 | | 10-13 | 8 (clone) |
| Tree position | base | top | base | top | industrial chips | |
| Basic density (kg/m$^3$) | 616 | 489 | 460 | 387 | 536 | 460 |
| Fibre length, weighted in length (mm) | 0.99 | 0.77 | 0.75 | 0.65 | 0.72 | 0.80 |
| Fibre width (μm) | 19.4 | 17.8 | n.a. | 17.9 | 16.5 | n.a. |
| Coarseness (mg/m) | 0.062 | 0.048 | 0.061 | 0.066 | 0.079 | 0.055 |
| Curl (%) | n.a. | n.a. | n.a. | 6.3 | 5.7 | n.a. |
| Number fibres/g | 19.7 x10$^6$ | 31.3 x10$^6$ | n.a. | n.a. | n.a. | n.a. |
| Reference | Anjos et al. 2011 | | Gil et al. 1999 | | Santos et al. 2006 | Gil et al. 1999 |

Table 1 resumes the results for the kraft cooking performance of *A. melanoxylon*, using *E. globulus* as a reference, and with values for other representative *Acacia* species as well. The results showed that *A. melanoxylon* grown in Portugal has a kraft pulping behaviour comparable with that of *E. globulus*, when submitted to the same cooking conditions (Table 1) (Gil et al. 1999; Santos et al. 2006; Santos et al. 2012a). The residual lignin (measured as kappa number) is of the same order and the pulp yield can be slightly higher for *A. melanoxylon*. The pulp viscosity, which measures the average degree of polymerization of polysaccharides, is very similar for the pulps of the two species, before and after bleaching. Considering that *E. globulus* wood is recognized as having an extraordinary performance in kraft pulping, requiring mild reaction conditions with low maximum temperature and moderate alkali charge, the results for *A. melanoxylon* are very promising.

Clark et al. (1991) and Guigan et al. (1991) reported pulp yields for several *Acacia* species in the range of 44.5–58.7% with kappa number in the range of 13.1-30.1 with a pulping selectivity (yield vs kappa number and pulp viscosity vs kappa number) at least as good as that obtained for *E. globulus*. Pan and Lu (1988), Logan and Balodis (1982) and Griffin et al. (2014) obtained pulp yields of *A. melanoxylon* higher than those reported before, namely 54.6%, 52.3% and 50.8% respectively (Table 1). The pulp yields obtained with *A. melanoxylon* are in the same range of those obtained with *A. mangium*, one of the most important fast growing species in Asia (Table 1).

The bleachability of the *A. melanoxylon* pulps was studied by Santos et al. (2006) using a $D_0E_1D_1E_2D_2$ sequence where the $D_0$ charge was 0.2% of the kappa number and the $D_1$ and $D_2$ of respectively 1.6% and 0.6%. The brightness obtained was 86.0% with a total chlorine dioxide consumption of 4.38% on Pulp. Santos et al. (2006) reported a similar behaviour between *E. globulus* industrial chips and *A. melanoxylon* and *A. dealbata* in a DEDED bleaching sequence if the kappa number of the unbleached pulp is taken in account for the chloride dioxide charge in the first stage.

Guigan et al. (1991) reported that a CEHD bleaching sequence of *A. mearnsii* and *A. silvestris* required more 10 kg active chlorine per ton of pulp than eucalypts, but other species, such as *A. mangium* and *A. auriculiformis* showed an oxidant consumption similar to eucalypt (Logan and Balodis 1982; Karim et al. 2011).

The quality of pulps is also determined by the physical and morphological properties of the pulp fibres. In particular, the intrinsic fibre strength (correlated with pulp viscosity) and the paper fibre structure are of high importance. The structure is mainly determined by morphological properties, such as fibre length, width and coarseness, including their distribution, and fines content, as well as by the technology and operating conditions used in the industrial paper formation process (Niskanen et al. 1998; Retulainen et al. 1998, Paavilainen 2000). Despite the strong influence of the paper technological solutions used, the pulp fibres have characteristics that make one pulp more suitable than another for a given paper product. The excellent performance of *E. globulus* kraft pulp for office paper is mainly due to the high number of fibres per gram, the relatively stiff fibres, and the moderate hemicellulose content (Santos et al. 2008, Paavilainen 2000). The number and stiffness of fibres give high light scattering and lead to high opacity, whereas both the relatively stiff fibres and the low hemicellulose content require fibre beating for external fibrillation to develop tensile strength. For instance, it is the conjugation of the different physical and morphological characteristics of the *E. globulus* kraft pulp fibres that enables good opacity at good mechanical strength, and good smoothness and dimensional stability (Pavilainen 2000).

Both fibre stiffness and collapsibility can be measured, but they are usually inferred by pulp fibre width and coarseness which are much easier to measure. Figure 1 shows that the wet fibre flexibility of *A. melanoxylon* (M-387) pulp fibres is higher than that from *E. globulus* (E-Ind) pulp fibres, which leads to higher paper densification with beating (Figure 2). These results reflect the morphological fibre properties, which are related with basic wood density. Higher basic wood density leads, in general, to lower wet fibre flexibility (Santos et al. 2008), which means that *A. melanoxylon* with higher basic wood density provide fibres with lower wet fibre flexibility, close to those of *E. globulus*.

For *A. melanoxylon* pulps produced from young trees, it seems that fibre length and coarseness of the pulp fibres are lower than those of the corresponding *E. globulus* pulp fibres (Table 2). With 40-year-old *A. melanoxylon* trees, the pulp fibre length is similar or higher than that of *E. globulus* industrial pulp fibres (Anjos et al. 2011). Fibre coarseness of *A. melanoxylon* pulps is in general somewhat lower than that of the *E. globulus* pulp fibres (Table 2). *A. melanoxylon* has wider fibres than *E. globulus*, which might promote excessive collapsibility for an office paper.

The wood basic density of *A. melanoxylon* has a large impact on the fibre morphological characteristics: wood samples with high basic density (514-616 kg/m$^3$) yield pulp fibres with comparable coarseness and fibre length to those of pulp fibres obtained from industrial *E. globulus* chips (536 kg/m$^3$) while samples with basic density in the range 449-505 kg/m$^3$ yield pulp fibres with lower coarseness but similar or higher fibre width (Santos et al. 2012a). Coarseness and fibre width induce collapsibility and fibre flexibility, and a dense paper structure may be obtained with low beating.

## 4. PAPER MAKING POTENTIAL

Although paper properties are affected by the technology used and the operating conditions in the different unit operations involved, the pulp characteristics play a significant role in the final paper properties and the overall process economy. To foresee the behaviour of a given pulp in paper production, the usual way is to submit the pulps at increasing levels of beating, produce paper handsheets and evaluate their optical, structural and mechanical properties. In fact, the development of the paper mechanical resistance requires the internal and external fibrillation of the pulp fibres given by beating that promotes interfibre hydrogen bonds and fibre entanglements, which both lead to structure cohesion, even in the wet state. The price to pay for this structural consolidation and mechanical resistance increase is the decrease of pulp suspension drainability, measured as an increase in the Schopper-Riegler degree (°SR).

Figure 2 shows the evolution of drainability of *A. melanoxylon* laboratorial bleached kraft pulps and of the reference *E. globulus* bleached kraft pulps (data from the same lab). Wood samples with different tree age and different basic wood density were selected. The experimental results clearly show that pulp fibres with higher coarseness (M-616, Table 1) i.e., higher fibre wall thickness at the same fibre width, develop drainage resistance much slower than the corresponding pulp fibres with lower coarseness (M-489; M-387, Table 2).

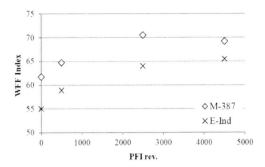

Figure 1. Wet fibre flexibility index as a function of PFI revolutions of *A. melanoxylon* (M-387) and *E. globulus* (E-Ind) pulps (Santos et al. 2006).

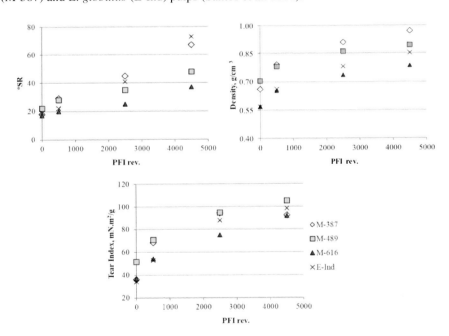

Figure 2. Schopler-Riegler degree, paper handsheet apparent density, tensile index as a function of PFI revolutions for pulps of *A. melanoxylon* (M-) with different wood basic density and of *E. globulus* (E-Ind) (Santos et al. 2006, Anjos et al. 2011, Guigan et al. 1991).

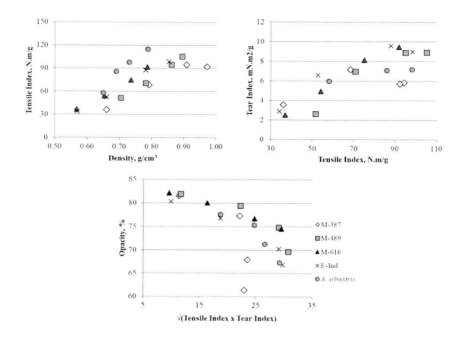

Figure 3. Tensile index, tear index and paper handsheet opacity as a function of paper handsheet apparent density, for pulps of *A. melanoxylon* (M-) with different wood basic density, of *E. globulus* (E-Ind) and *A. silvestris* (Santos et al. 2006, Anjos et al. 2011; Guigan et al. 1991).

A bleached kraft pulp produced from industrial *E. globulus* chips exhibits an intermediate behaviour for the low range of beating and surpass the *A. melanoxylon* samples at high beating levels. This is probably due to the abundant external fibrillation shown by *E. globulus* (relatively high coarseness) (Santos et al. 2008). Literature data (Guigan et al. 1991) of bleached kraft pulp from *A. silvestris* grown in Australia (basic wood density: 551 kg/m3) are also included for comparison. The development of the paper handsheets apparent density is also shown is Figure 2. The paper sheet density of *E. globulus* is relatively low, and it is possible to select an *A. melanoxylon* wood sample with similar behaviour. Noteworthy is the extremely low apparent paper density provided by the M-616 pulp. According to the data, the wood should have a basic density in the range between 489 and 616 kg/m$^3$, which match the basic density of the *E. globulus* selected as a reference in this study (536 kg/m$^3$). As expected, tensile index (Figure 2) increases significantly with beating, but the results depend both on species and basic wood density.

The tensile index at a given apparent paper density (Figure 3) is an important quality parameter and some differences can be observed; the M-387 pulp has significantly lower tensile strength than the M-616 pulp, at a given handsheet density, but the M-616 pulp requires more energy in beating to attain a given density (Figure 2). However, *A. melanoxylon* wood samples with intermediate density can be selected to provide pulps with tensile index/sheet density relationships comparable with *E. globulus*. *A. silvestris* exhibits the highest performance.

Tear index vs tensile index (Figure 3) is another important relationship to consider in the papermaking potential evaluation. The performance of *Acacia* is very close although slightly lower than that of *E. globulus*. This is a consequence of the fibre dimensions conjugated with paper density, which determine both tensile and tear index.

Another critical functional property in office papers is opacity and its relation with apparent density and mechanical strength. Figure 3 shows the opacity as a function of handsheet mechanical strength (measured as square root of the product of tensile index and tear index) and similar performance can be noticed between *Acacia* and *E. globulus*, with the exception of the M-387 with extremely low basic wood density. This last pulp has the lowest average fibre length (0.65 mm; Table 2), which negatively affects tensile index and particularly tear index, despite the strong structure densification with beating; this behaviour leads to low opacity for a given mechanical performance. The *Acacia* wood samples with higher basic wood density perform similarly to *E. globulus*

Another important paper property is smoothness. Fibres with lower coarseness, which is found in fibres with low wall thickness, tend to collapse easier and provide paper with higher smoothness. However, at the same apparent paper density, the disadvantage of coarse fibres is very small (Santos 2006).

## 5. PREDICTION OF PULP QUALITY FEATURES

Writing and printing papers have demanding requirements e.g., regarding density, mechanical strength, optical and surface properties that depend on the overall pulp features. For the final product performance, the pulping and bleaching conditions are important as well as the characteristics of the initial fibre source. Since the pulp feedstock for writing and printing papers is made up of 75-80% of fibres, the remaining being different additives, it is clear that the paper properties and its in-use quality will strongly depend on the fibres morphology and strength.

The measurement of paper properties is time consuming and costly, and therefore their prediction using a few predictor variables is a seductive option, namely when prospective or screening studies are the focus. The use of machine learning methods or of spectroscopic approaches to properties assessment have therefore been researched.

Several studies were found using support vector machine regression techniques for predicting some mechanical properties of different materials, namely of wood and wood composites. For instance, prediction of fracture toughness of wood (Samarasinghe and Kulasiri 2007), MOR and MOE of particleboard (Fernández et al. 2008), some mechanical properties of solid wood and wood composites (Cook and Chiu 1997) as well as wood veneer classification (Castellani and Rowlands 2009). However, only a few works were found for paper properties prediction.

Anjos et al. (2015) applied different mathematical models to the prediction of paper mechanical and optical properties using paper density as the predictor variable for hardwood pulps of *E. globulus*, *A. melanoxylon* and *A. dealbata* with different refining levels. Multivariate regression techniques were applied to obtain straightforward prediction models that closely adhered to the available data and were valid when applied to independent data predictions. The goodness of fit was assessed by the coefficient of determination ($R^2$) and the root mean squared error (RMSE), and the dataset variability was measured through the total sum of squares (SStot), the regression sum of squares (SSreg) and the residual sum of squares (SSerr). They concluded that it was possible to have prediction models of paper properties for each of the hardwood species using unsupervised classification techniques and multivariable regression techniques, and that density had the highest explanatory power as predictor variable, depending on the refining level.

García-Gonzalo et al. (2016) performed a similar study for softwood species e.g., *Pinus pinaster and P. sylvestris*, *Cupressus lusitanica, C. sempervirens* and *C. arizonica,* and demonstrated that it is also possible to obtain good models with high coefficient of determination with two variables: the numerical variable density and the categorical variable species. These two works show the potential of such mathematical tools for prediction of pulp and paper properties.

Other techniques could be used to predict paper properties, namely using indirect assessments that can be correlated to specific properties. Infrared spectroscopy (IR) is one of the most important and powerful methods providing molecular vibrational spectral information that aids to identification of chemical functional groups and to properties associated to the chemical structure of the materials. IR has been applied in many fields, including agronomy, food science or materials engineering, and it is also an important tool for studying paper structure and chemistry (Pereira et al. 2016; Santos el al. 2012b, 2016; Trafela et al. 2007). Fourier-transform infrared (FTIR) and near infrared (NIR) spectroscopy are the most used techniques for pulp and paper characterization and quality control, including on-line process control. The use of FTIR and NIR spectral data information requires chemometric techniques to extract relevant information in relation to the specific property.

Raman spectroscopy has also been used for estimation of the composition of different pulp qualities and of residual lignin (Kuptsov 1994), for compositional mapping of papers and paper coatings (Workman Jr. 2011), paper analysis (Halttunen et al. 2001), the nature of photo-yellowing of lignin-containing pulps and papers (Umesh 2008), among other applications. Wójciak et al. (2014) assessed the influence of different treatment factors (e.g., pH change of the bleaching slurry from acidic to alkaline, presence of catalyst and stabilizers) on the pulp and its constituents by comparing NIR, FTIR and FT-Raman spectral responses. They found that the three techniques are complementary since they identify different properties: 1) NIR spectroscopy detects differences in oxygen-delignified kraft pulps bleached with different processes and can estimate pulp optical parameters or ISO brightness values; 2) FT-Raman spectroscopy allows differentiating the oxygen-delignified pulps bleached with hydrogen peroxide under acidic and alkaline conditions but no correlations were found for optical properties; 3) FTIR is sensitive to the specific bleaching conditions of oxygen-delignified pulps, but the results are not clear for monitoring of pulps with very low lignin content.

The application of these techniques as pulp and paper prediction tools for different kinds of raw material from *Acacia* species has been researched mostly by our research team, especially for *A. melanoxylon* (Pereira et al. 2016; Santos et al. 2016, 2015, 2014). These studies have confirmed the applicability of NIR-PLSR models for prediction of different properties relevant for pulp and paper industry for *A. melanoxylon*.

Santos et al. (2016) predicted kraft pulp yield using near-infrared (NIR) spectroscopy under identical pulping conditions and a kappa number of 15 with very good correlations. In this study PLS regression modelling used a total of 75 pulp samples of *A. melanoxylon* with a pulp yield variation range of 47.0–58.2% by using cross-validation with 62 samples and an external validation with 13 samples. The best model used a 2ndDer pre-processing in wavenumber ranges from 9087 to 5440 and 4605 to 4243 cm$^{-1}$ with a standard error of prediction of 0.4%, a $R^2$ of 98.1%, and RPD of 4.8.

Santos et al. (2015) studied the ability of NIR spectroscopy for predicting the ISO brightness of *A. melanoxylon* unbleached kraft pulps. A PLSR model was developed based on 75 pulp samples with a brightness variation range of 18.9 to 47.9% using 1stDer pre-processed spectra in wavenumber ranges from 9404 to 7498 cm$^{-1}$ and 4605 to 4243 cm$^{-1}$ with a root mean square error (RMSEP) of brightness prediction of 0.5%, a $R^2$ of 99.5% and a very robust RPD (ratio of performance to deviation) of 14.7.

Santos et al. (2014) developed a NIR–PLS-R model for Kappa number prediction with 75 pulp samples of *A. melanoxylon* with a narrow Kappa number variation range of 10 to 17. Very good correlations models were obtained with 1stDer pre-processing in the wavenumber range from 6110 to

5440 cm$^{-1}$ with a root mean square error of prediction of 0.4 units of Kappa number, a R$^2$ of 92.1%, and RPD of 3.6.

Pereira et al. (2016) applied NIR with PLS-R models based on spectral data obtained for the wood to predict morphological properties of fibres (length and width) of unbleached kraft pulp of *A. melanoxylon*. The results showed a high accuracy and precision, with RPD of 3.9 and 3.3. The selected models for fibre length and fibre width used the second derivative and first derivative + multiplicative scatter correction pre-processing, in the wavenumber ranges from 7506 to 5440 cm$^{-1}$ and obtained the statistical parameters of cross-validation (RMSECV of 0.009 mm and 0.39 mm) and validation (RMSEP of 0.007 mm and 0.36 mm) with RPDTS values of 3.9 and 3.3, respectively.

## CONCLUSION

The available data on the *A. melanoxylon* wood features of relevance as quality parameters for pulping e.g., fibre biometric parameters, chemical composition, delignification behavior, and the potential for paper making e.g., paper physical and optical properties showed that the species can be considered as a fibre raw-material for production of short fibre pulps comparable to those of *Eucalyptus globulus*. Similar influence of heartwood in the feedstock apply as shown by the studies both on *A. melanoxylon* (Lourenço et al. 2008) and *E. globulus* (Lourenço et al. 2010).

*A. melanoxylon* has a rapid growth under favourable conditions and it is foreseeable that under the same environmental conditions it will have similar annual increments and productivities as *E. globulus*. Two management regimes could be potentially foreseen for *A. melanoxylon*: i) an intensive short rotation plantation directed for pulpwood production, similar to what is presently carried out in the commercial *E. globulus* plantations (with 9-13 years rotation); or ii) a plantation directed for sawmilling quality products with rotation above 40 years, with the young trees obtained from thinning operations to be directed to pulping.

In the regions where *A. melanoxylon* is an invasive, the wood availability will be given by the erradication of the spontaneously invading plants. The use of such wood material as a pulping raw material may therefore contribute to partially support the costs of the control programmes.

In summary, it seems possible to further explore the possibility of using *A. melanoxylon* wood to produce papers with mechanical and optical properties suitable for quality printing papers. This goes in line with the fact that diversification of industrial fibre sources is a matter of research with the characterization of different raw materials in view of their pulping potential.

## ACKNOWLEDGMENTS

Centro de Estudos Florestais is a research unit funded by Fundacão para a Ciência e a Tecnologia (FCT) within UID/AGR/UI00239/2013. Ana Lourenço acknowledges a post-doc fellowship from FCT, SFRH/BPD/95385/2013. Support from FibEnTech - Fiber Materials and Environmental Research Unit is also acknowledged.

## REFERENCES

Anjos, O; García-Gonzalo, E; Santos, AJA; Simões, R; Martínez-Torres, J; Pereira, H; García-Nieto, PJ. *BioResources.*, 2015, 10(3), 5920-5931.

Anjos, O; Santos, A; Simões, R. *Appita Journal.*, 2011, 64(2), 185-191.

Bektas, I; Tutus, A; Eroglu, H. *Turk J Agri For.*, 1999, 23 (3), 589-597.

Castellani, M; Rowlands, H. *Eng Appl Artif Intel.*, 2009, 22, 732–41.

Clark, NB; Balodis, V; Guigan, F; Jingxia, W. *Proc. Int. Workshop on Advances in Tropical Acacia Research*, Bangkok. 1991, pp. 138.

Coleman, MJ. *Tappi J.* 1998, 81 (12), 43-49.

Cook, DF; Chiu, CC. *Eng Appl Artif Intell.*, 1997, 10(2), 171–177.

El-Hosseiny, F; Anderson, D. *Tappi J.*, 1999, 82, 202–20.

Fernández, FG; Esteban, LG; DePalacios, P; Navarro, N; Conde, M. *Invest Agrar – Sist R.*, 2008, 17, 178–87.

Furtado, JLF; Peralba, MCR; Zini, CA; Caramão, EB. *Holzforschung.*, 2000, 54, 159-164.

García-Gonzalo, E; Santos, AJA; Martínez-Torres, J; Pereira, H; Simões, R; García-Nieto, PJ; Anjos, O. *BioResources.*, 2016, 11(1), 1892-1904.

Gil, C; Amaral, ME; Tavares, M; Simões, R. *Proc. 1º Encontro sobre Invasoras Lenhosas [Proc. 1st Meeting on Woody Invasives]*, Gerês, Portugal, 1999, pp 171.

Griffin, A; Twayi, H; Braunstein, R; Downes, G; Son, D; Harwood, C. *Appita J.*, 2014, 67(1), 43-49.

Griffin, AR; Midgley, SJ; Bush, D; Cunningham, PJ; Rinauto, AT. *Divers Distrib.* 2011, 17, 837-847.

Guigan, F; Balodis, V; Jingxia, W; Clark, NB. Proc. *Int. Workshop on Advances in Tropical Acacia Research*, Bangkok. 1991, pp145.

Halttunen, M; Vyörykkä, J; Hortling, B; Tamminen, T; Batchelder, D; Zimmermann, A; Vuorinen, T. *Holzforschung.*, 2001, 55(6), 631-638.

Hillis, WE. *Heartwood and Tree Exsudates*; Springer Series in Wood Science; Springer Verlag: Syracuse, NY, 1987; pp 239.

Igartúa, DV; Moreno, K; Monteoliva, SE. *Forest Syst.*, 2017, 26 (1), p. e007

Jahan, SM; Sabina, R; Rubaiyat, A. *Turk J Agric For.*, 2008, 32, 339-347.

Karim, MR; Islam, MN; Malinen, RO. *Wood Sci Technol.*, 2011, 45, 473-485.

Knapic, S; Tavares, F; Pereira, H. *Forestry.*, 2006, 79, 371–380.

Kuptsov, AH. *Vib Spectrosc.*, 1994, 7(2), 185-190.

Logan, AF; Balodis V. *The Malaysian Forester.* 1982, 42, 217-236.

Lourenço, A; Gominho, J; Pereira, H. *J Pulp Pap Sci.*, 2010, 36(3), 85.

Lourenço, A; Baptista, I; Gominho, J; Pereira, H. *J Wood Sci.*, 2008, 54(6), 464-469.

Malan, FS; Male, JR; Venter, JS. *Paper South Africa*, 1994, 2, 6–16

Miranda, I; Gominho, J; Lourenço, A; Pereira, H. *Appita J.*, 2007, 60, 485–500.

Nicholas, I; Brown, I. *Blackwood, a Handbook for Growers and Users*. New Zealand: Rotorua; 2002.

Niskanen, K. *Paper Physics* (*Papermaking Science and Technology*); Tappi Pr, Fapet Oy, Helsinki, Finland, 1998, pp 342.

Paavilainen L. *Paperi Ja Puu*. [*Paper And Wood.*], 2000, 82(3), 156-161

Pan, Z; Lu, X. *Proc. of the Use of Australian trees in China Workshop*. Guangzhoou, China 1998.

Pedieu, R; Riedel, B; Pichette A. *Eur J Wood Wood Prod.*, 2009, 67(1), 95-101.

Pereira, H; Santos, AJA; Anjos, O. *Materials*, 2016, 9(1), 8.

Retulainen, E; Niskanen, K; Nilsen, N. *Paper Physics* (*Fibers and Bonds*); Tappi Pr, Fapet Oy, Helsinki, Finland, 1998, pp 54-87.

Samarasinghe, S; Kulasiri, D; Jamieson, T. *Silva Fenn.*, 2007, 41, 105–22.

Santos, AJA; Anjos, O; Pereira, H. *Wood Sci Technol.*, 2016, 50(6), 1307-1322.

Santos, A; Anjos, O; Pereira, H. *Forest Syst.*, 2015, 24(2), eRC03, 6 pages.

Santos, AJA; Anjos, O; Simões, R; Rodrigues, J; Pereira, H. *BioResources.*, 2014, 9(4), 6735-6744.

Santos, A; Anjos, O; Amaral, ME; Gil, N; Pereira, H; Simões, R. *J Wood Sci.*, 2012a, 58, 479–486.

Santos, AJA; Alves, AMM; Simões, RMS; Pereira, H; Rodrigues J; Schwanninger M. *J Near Infrared Spec.*, 2012b, 20(2), 267-274.

Santos, A; Amaral, ME; Vaz, A; Anjos, O; Simoes, R. *Tappi J.*, 2008, 7(5), 25–32.

Santos, A; Anjos, O; Simões, R. *Appita J.*, 2006, 59 (1), 58-64.

Searle, SD. *Aust Forestry.*, 2000, 63(2), 79-85.

Silva, JC; Borralho, NMG; Araújo, JÁ; Vaillancourt, RE; Potts, BM. *Tree Genet Genomes.*, 2009, 5, 291–305.

Trafela, T; Strlič, M; Kolar, J; Lichtblau, DA; Anders, M; Mencigar, DP; Pihla, B. *Anal Chem.*, 2007, 79(16), 6319-6323.

Umesh PA. *J Wood Chem Technol.*, 2008, 18(4), 381-401.

Valente, CA; Mendes de Sousa, AP; Furtado, FP. Carvalho, AP. *Appita J.*, 1992, 45(6), 403-407.

Watanabe, K; Miyanishi, T. *Japan Tappi J.*, 2004, 58(8), 1097-1103.

Wójciak, A; Kasprzyk, H; Sikorska, E; Krawczyk, A; Sikorski, M; Wesełucha-Birczyńska, A. *Vib Spectrosc*, 2014, 71, 62-69.

Workman Jr., JJ. *Appl Spectrosc Rev.*, 2011, 36, 139-168.

# INDEX

## A

*A. acanthoclada*, 40
*A. aneura*, 32
*A. colletioides*, 40
*A. dealbata*, 11, 13, 16, 34, 50, 242, 248
*A. ferocior*, 40
*A. longifolia*, 15, 16, 50, 51, 55
*A. mangium*, vii, x, 11, 12, 13, 16, 17, 34, 84, 85, 87, 89, 90, 92, 93, 95, 96, 97, 98, 237, 240, 242
*A. mearnsii*, 12, 16, 17, 50, 51, 57, 236, 237, 242
*A. melanoxylon*, vii, viii, ix, xv, 2, 3, 4, 5, 6, 7, 8, 9, 10, 12, 13, 14, 15, 16, 17, 18, 19, 34, 236, 237, 238, 239, 240, 241, 242, 243, 244, 245, 246, 247, 248, 250, 251, 252
*A. pycnantha*, 55
*A. saligna*, 50
*A. spinescens*, 40
*A. stricta*, 55
Aca f 1, xiii, 194, 199, 200, 201, 204, 205, 210
Aca f 2, xiii, 194, 201, 202, 205, 206, 211
*Acacia aptaneura*, 48, 69
*Acacia ehrenbergiana*, 32
*Acacia farnesiana*, xiii, 66, 194, 195, 197, 204, 210, 211
*Acacia ligulata*, 45
*Acacia melanoxylon*, v, viii, xiv, 1, 2, 11, 18, 20, 197, 208, 236, 241
*Acacia s.l.*, vii, x, 28, 29, 30, 31, 32, 34, 35, 39, 43, 45, 46, 48, 49, 50, 52, 53, 57, 58, 187, 217, 218, 227, 231
*Acacia s.s.*, 30, 53
*Acacia suaveolens*, 48
*Acacia-Pseudomyrmex* mutualism, viii, xiv, 214, 223
Acacieae, v, ix, 27, 28, 29, 31, 32, 33, 34, 35, 38, 40, 41, 42, 43, 44, 50, 53, 55, 57, 58, 59, 208
*Acaciella*, ix, 27, 28, 30, 31
acid, 11, 12, 13, 14, 44, 99, 114, 115, 116, 118, 120, 122, 123, 124, 125, 127, 128, 129, 130, 132, 133, 137, 140, 141, 142, 143, 144, 145, 146, 147, 148, 149, 150, 151, 152, 159, 202, 204, 205, 206
acidic, 12, 35, 36, 142, 150, 158, 249

# Index

acidity, 147
Africa, ix, xii, 3, 5, 18, 20, 27, 28, 29, 30, 31, 35, 41, 44, 50, 52, 57, 60, 63, 66, 180, 182, 193, 196, 218, 228, 231, 236, 253
agroforestry systems, 36, 64, 66
alien species, 52, 54
alkaloids, 13, 166, 179, 216, 219
allergens, viii, xiii, 194, 195, 197, 199, 201, 202, 203, 206, 208, 209, 210, 212
allergic reaction, 195
allergic rhinitis, viii, xii, 193, 195, 197, 207
allergy, viii, xii, 13, 193, 194, 195, 196, 206, 207, 208, 209, 210, 211, 212
*Aloysia gratissima*, 46
America, ix, xii, 3, 27, 28, 29, 30, 31, 74, 76, 77, 101, 102, 181, 182, 187, 189, 193, 196, 218, 220, 230, 231
amino acid, xiii, 124, 138, 139, 140, 142, 163, 166, 194, 199, 200, 201, 202, 204, 205, 212, 216, 219
anatomy, ix, 2, 106, 187, 238
ancestral character states, viii, xi, 161, 163, 164, 167, 185, 188, 189
ants, viii, xi, xiii, 29, 40, 41, 42, 43, 44, 56, 59, 60, 69, 162, 165, 166, 167, 174, 175, 176, 177, 178, 179, 182, 183, 187, 213, 215, 216, 217, 218, 219, 220, 221, 222, 223, 224, 225, 226, 227, 228, 229, 230, 231, 232, 233
*Apis mellifera*, 57
arabinogalactan, 116, 133, 137
Argentina, 3, 27, 32, 34, 37, 43, 46, 47, 49, 50, 53, 55, 59, 60, 63, 64, 66, 70, 71, 73, 74, 75, 76, 77, 78, 79, 80
arid lands, 30, 35, 114
Asia, xii, 6, 24, 44, 100, 103, 108, 111, 193, 196, 207, 242
Asian countries, 110
Aspidosperma quebracho-blanco, 38, 54
asthma, 197, 207, 208

Australia, viii, ix, xii, 2, 7, 8, 10, 15, 21, 25, 27, 28, 30, 31, 33, 35, 41, 46, 48, 52, 53, 60, 61, 63, 64, 65, 66, 69, 193, 196, 236, 246

## B

bacteria, 29, 59, 67
biodiversity, xiv, 35, 37, 63, 66, 84, 99, 106, 235
bioenergy, 19
biogeography, 60, 231
biological control, 54, 56, 65
biomass, 22, 37, 46, 47, 48, 51, 52, 54, 99, 237
biosynthesis, 4
biotic, xiv, 32, 50, 52, 228, 236
birds, 45, 49, 59, 61, 219, 230
bleaching, 238, 242, 248, 249
Botswana, 41, 60, 196, 209
Brazil, 30, 33, 60, 62, 99, 101, 110, 236
bronchial asthma, 195

## C

carbon, 36, 61, 64, 97, 121, 123, 125, 131, 132, 142
cellulose, 12, 237, 239
Cenchrus ciliaris, 46
Chaco Árido, 34, 68, 71, 75, 77
Chaco Serrano, 34
chemical, vii, ix, xiv, 2, 4, 10, 11, 15, 18, 62, 114, 117, 119, 120, 121, 122, 126, 128, 130, 133, 136, 140, 141, 143, 144, 145, 146, 150, 153, 155, 166, 179, 186, 216, 219, 220, 236, 238, 239, 249, 251
chemical characteristics, xiv, 219, 236
chemical composition, vii, ix, 2, 4, 10, 11, 126, 238, 239, 251
chemical properties, 18
chemometric techniques, 249

# Index

*Chenopodium album*, 200, 203, 204, 205, 210, 211
Chile, 3, 10, 34, 53, 55, 56, 66
China, 20, 69, 102, 103, 106, 108, 254
Chubut, 34
classic methodology for carbohydrates, viii, xi, 113, 114, 117
clayey soils, 32, 35
climate change, 50, 54, 207
$CO_2$, xiv, 31, 60, 182, 213, 217
coastal ecosystems, 35
coastal region, 219
colonization, 55, 68, 97, 99, 187, 227, 228
communities, 35, 45, 46, 50, 56, 57, 64, 67, 71, 99, 181
comparative method, 163, 186
competition, 37, 38, 44, 47, 52, 56, 57, 58, 62, 65, 66, 68, 95, 96
competitive advantage, 58, 175
competitors, 36, 37, 53, 58, 61
component-resolved diagnosis, viii, xii, 194, 197
compounds, 4, 10, 12, 16, 40, 53, 120, 179, 186, 219, 220, 230
contact dermatitis, 18, 197, 208
Coomassie Brilliant Blue staining, 198
Costa Rica, 75, 110, 184, 219, 221, 225
*Crematogaster mimosae*, 38
Cretaceous, 31, 180, 186
cross-reactivity, xiii, 194, 199, 202, 203, 206, 208, 210, 211, 212
cross-validation, 250, 251

## D

decomposition, xiv, 61, 213, 217
defense mechanisms, 216, 219
degenerated primers, 201
degradation, 85, 95, 110, 111, 117, 121, 125, 128, 138, 155
degraded area, 85

desert, 35, 62, 67, 69, 207, 208
discrete variable, 167, 169
diseases, viii, xii, 194, 196
distribution, viii, xi, xiv, 2, 13, 29, 30, 31, 32, 34, 35, 44, 64, 66, 67, 107, 161, 162, 164, 183, 185, 214, 217, 219, 220, 221, 223, 225, 226, 227, 229, 232, 243
divergence, 163, 164, 166, 186, 227, 229
diversification, 32, 35, 52, 164, 165, 181, 186, 188, 214, 219, 229, 252
diversity, x, 53, 63, 84, 86, 88, 93, 94,

## E

ecological processes, x, 52, 84, 97
ecological systems, 45
ecology, 59, 68, 75, 85, 103, 106, 108, 185, 187, 188, 228, 231
ecosystem, 37, 51, 53, 56, 65, 85, 108
elucidation, 118, 119, 144, 146, 155, 156, 157, 197
energy, xv, 19, 44, 176, 178, 236, 247
energy consumption, xv, 236
energy expenditure, 176, 178
environmental characteristics, 36
environmental conditions, 41, 114, 174, 251
environments, 30, 42, 46, 51, 52, 58, 59, 68, 95, 96, 121, 124, 125, 133, 136, 141, 143, 147, 148, 183, 184, 215
enzyme, 95, 101, 179, 221
epitopes, 199, 206, 211
Espinal, 34, 70, 71
ethanol, 10, 11, 13
Europe, v, viii, 1, 2, 3, 65, 206
evapotranspiration, 37, 54, 57
evolution, viii, xi, 28, 29, 30, 50, 52, 53, 64, 65, 69, 103, 161, 162, 163, 164, 165, 170, 176, 177, 178, 179, 180, 182, 183, 184, 185, 186, 187, 188, 189, 190, 217, 227, 228, 229, 230, 231, 233, 244
exotic plantation, 84, 97

experimental condition, xv, 236
extractives, ix, xv, 2, 4, 10, 11, 12, 13, 15, 16, 236, 238, 239
extracts, xii, 13, 16, 17, 114, 194, 197, 201, 203, 207
exudate, 115, 116, 122, 123, 124, 126, 127, 128, 129, 131, 132, 134, 135, 136, 138, 144, 150, 151, 152, 154, 156, 157, 158

## F

Fabaceae family, xii, 193, 196, 203, 205
families, viii, xi, 12, 113, 116, 206, 215
fauna, vii, x, 21, 28, 29, 46
fertility, x, 39, 42, 84, 97
fibres, xv, 6, 9, 10, 236, 241, 242, 243, 244, 247, 248, 251
fibres morphology, 236, 248
fire, x, 3, 27, 28, 29, 31, 32, 40, 41, 42, 45, 46, 47, 48, 49, 50, 51, 52, 53, 54, 56, 58, 59, 60, 61, 63, 64, 65, 66, 68, 69, 72, 75
fire cycles, 40
fire event, x, 27, 29, 46, 48, 49, 50, 51, 53, 56, 58, 59
Florentine Valley, 34, 62
food, xiii, 35, 41, 42, 166, 177, 180, 182, 186, 213, 216, 217, 219, 230, 231, 249
forest ecosystem, 110
forest fire, x, xiv, 27, 29, 46, 47, 61, 236
forest restoration, 95, 106, 107
forest structure, 84, 85, 97
*Fraxinus excelsior*, 195, 200, 204, 205, 210
fruits, xiii, 43, 49, 176, 186, 194, 201
fungi, xiii, 18, 55, 59, 114, 207, 213, 217
fynbos, 51, 54, 56, 57, 61, 63

## G

genus, ix, 2, 13, 27, 28, 30, 31, 38, 40, 49, 62, 65, 120, 124, 150, 154, 165, 173, 174, 188, 190, 202, 203, 212, 218, 225, 226, 227, 229
geographic mosaic theory, vi, 213
geological history, 32
germination, x, 21, 27, 43, 44, 45, 48, 49, 51, 64, 68, 69, 95, 210
glucose, 12, 114, 176, 222
glycoproteins, 116, 197
glycosylation, 197, 201, 202
Glyptodontidae, 40
Gomphoteridae, 40
Gondwanaland, 31
gracilis, 225, 226
Gran Canaria Island, 45
grasses, xiii, 31, 32, 37, 38, 41, 46, 47, 48, 49, 50, 53, 65, 194, 195, 201
grasslands, vii, x, 32, 46, 49, 50, 52, 60, 66, 79, 84, 85, 86, 87, 94, 97, 98, 177, 180, 181, 182
grazing, 40, 44, 46, 48, 50
growth, viii, ix, 2, 3, 4, 6, 17, 32, 34, 37, 38, 39, 41, 42, 49, 50, 51, 54, 55, 61, 62, 63, 69, 97, 99, 101, 107, 174, 175, 182, 215, 251
growth rate, 4, 6, 39, 42, 97, 175, 182
growth rings, ix, 2, 6, 17, 60
gum Arabic, 158, 159, 197, 208
gum exudates, vii, xi, 113, 114, 116, 117, 119, 120, 122, 137, 147, 153, 154, 156, 157, 158

## H

habitats, x, 28, 29, 30, 32, 42, 46, 50, 57, 58, 95, 173, 174, 175, 181, 182, 186
hardwoods, 18, 236
hay fever, viii, xii, 193, 195
heartwood, viii, ix, xv, 2, 4, 5, 6, 7, 10, 11, 12, 13, 14, 18, 19, 20, 21, 236, 237, 238, 239, 251, 253

height, 3, 5, 6, 14, 41, 44, 46, 49, 86, 95, 96, 169, 171, 172, 173, 183
heterogeneous systems, 116
heteropolysaccharides, xi, 113, 114, 115, 116, 117, 120, 142, 144, 150, 154
history, 52, 58, 60, 66, 109, 162, 164, 166, 216, 230
humidity, 96, 175, 177, 182
hydrolysis, 118, 120, 133, 137, 230

## I

IgE-biding reactivity, 198
IgE-binding protein, 201
IgE-binding site, 199, 203, 206
III and the VIII Region, 34
immunoblotting assay, 198
immunochemical, xiii, 194, 197, 208, 210
immunotherapy, viii, xii, xiii, 194, 197, 203, 209, 212
India, 3, 36, 64, 103, 136, 209
individuals, 179, 182, 195, 222, 226
Indonesia, 33, 101, 104, 108, 196, 209, 236, 237
industries, xv, 236, 237
industry, vii, ix, xiv, xv, 2, 157, 235, 236, 250
infrared spectroscopy, 249
insects, 42, 57, 114, 176, 179, 182, 216, 217, 219, 221, 222, 230
invasive, 3, 19, 21, 22, 29, 51, 52, 53, 54, 55, 56, 57, 58, 61, 63, 64, 238, 252
invasiveness, 28, 51, 52, 54, 56
Iran, viii, xii, 193, 194, 196, 206, 207, 211, 212
islands, 35, 50, 61
I-TASSER, 200, 201

## K

Kalahari Desert, 36, 38

kappa number, viii, xv, 236, 237, 238, 240, 241, 242, 250
Korea, 83, 104, 107, 108, 109, 110
Kraft pulp, 236

## L

landscape, xi, 35, 84, 85, 98, 102, 111, 181, 183
larvae, 166, 177, 219, 222
legume, 52, 60, 64, 97
leucine, 123, 135
light scattering, xv, 236, 243
lignin, ix, 2, 10, 11, 12, 22, 238, 239, 241, 249, 250
Ligustrum vulgare, 200, 204, 205
lipids, 166, 217, 219
*Lithraea molleoides*, 46
livestock, 35, 37, 39, 40, 41, 43, 46, 48, 54, 62, 72
*Lolium perenne*, xiii, 194, 203

## M

machine learning, 248
Machrauchenidae, 40
Macropus, 40
Malaysia, 104, 108, 196, 236
mammals, 40, 56, 60, 66
*Mariosousa*, ix, 27, 28, 30
mathematical models, 248
mechanical properties, vii, ix, 2, 15, 20, 238, 244, 248
Mediterranean, 2, 22, 208, 209
Megatheriidae, 40
Mesoamerica, 42, 219
metabolites, 29, 67, 68, 169, 170, 172, 178
metals, 128, 130, 142, 145, 151
methanol, 118, 237, 241
methodology, viii, xi, 113, 114, 117
methylation, 122, 141, 146, 153, 155

Mexico, 30, 34, 103, 181, 185, 189, 190, 191, 213, 219, 220, 221, 225, 229, 230, 231, 232, 234
microclimate, x, 84, 95, 98, 99, 101
microorganisms, vii, x, 28, 29, 43
Middle East, 44, 45, 67, 207
Mimosoideae, 2, 52, 60, 62, 68, 178, 187, 188, 189, 208, 229, 231, 232
Miocene, xii, 31, 35, 42, 64, 162, 166, 169, 180, 181, 182, 184, 219, 228
moisture, 18, 35, 36, 38
moisture content, 18
Mojave Desert, 34
molecular biology, 197
molecular structure, 115, 120, 199
molecular weight, 125, 198, 199, 201, 202, 203
molecules, xii, 116, 194, 195, 197, 203, 206
monomers, 12, 117, 118, 120
monosaccharide, 12, 119
Monte, 34, 56, 71, 75, 80, 81
morphology, 236, 238, 248
mortality, x, 48, 84, 88, 92, 93
mortality rate, x, 84, 88, 92, 93
mosaic, xiv, 214, 225, 227
mucus, xii, 194, 195, 197
mutualism, vi, xi, xiii, 38, 41, 42, 43, 66, 67, 68, 69, 162, 165, 179, 184, 186, 187, 213, 219, 220, 226, 227, 228, 229, 230, 231, 232, 233
Myanmar, 108
mycorrhizal, 55, 59
Mylodontidae, 40
myrmecochores, 44
Myrmecophily, 165, 215, 217

## N

native species, x, 28, 29, 51, 52, 54, 55, 56, 57, 84, 88, 95, 97

natural selection, viii, xiv, 214, 222, 223, 233
*Nauclea diderrichii*, 39
New Zealand, 3, 15, 21, 65, 253
N-glycosylation site, 201
NIR spectra, 249
NIR spectral data, 249
nitrogen, vii, x, xiv, 3, 19, 28, 29, 35, 36, 42, 46, 53, 55, 59, 64, 96, 97, 99, 100, 102, 213, 217, 229, 232
nitrogen fixation, 59
NMR spectroscopy, vii, xi, 113, 114, 116, 117, 122, 125, 138, 144, 146, 153, 154, 155
non-polar, 15
non-reducing SDS-PAGE, 198
Nototherium, 40
nucleus, 115, 116, 120, 121, 125, 128, 130, 132, 133, 135, 136, 137, 138, 140, 142, 145, 150, 153

## O

Ole e 1-like protein, xiii, 194, 199, 200, 204, 205, 210
*Olea europaea*, 195, 200, 202, 204, 205, 206, 210, 212
oligosaccharide, 117, 125, 133
open reading frame, 200, 201
optical properties, 248, 250, 251, 252
organic matter, 36, 101
Oryza sativa, 36
oxidation, 12, 118, 121, 122, 130, 133, 137, 151
Oxisols, 35

## P

Paleocene, 31
paleontology, 163
paper density, 246, 247, 248

# Index

paper properties, viii, xv, 4, 18, 236, 237, 244, 248, 249
papermaking potential, 236, 247
*Parasenegalia*, ix, 27, 28, 30, 68
parenchyma, ix, 2, 4, 8, 10, 44
permission, iv, 45, 121, 122, 123, 124, 125, 126, 127, 129, 130, 131, 132, 134, 135, 136, 137, 139, 140, 141, 143, 144, 146, 148, 149, 151, 152
personal communication, 226
phenolic compounds, 13
Philippines, v, vii, x, 83, 84, 85, 86, 93, 95, 98, 99, 100, 103, 104, 105, 106, 108
phylogenetic tree, 162, 164, 167, 169
physical and mechanical properties, 238
plants, xi, xiii, 22, 24, 29, 38, 39, 40, 41, 42, 49, 51, 52, 56, 59, 62, 64, 66, 68, 102, 103, 106, 162, 165, 166, 173, 174, 177, 178, 179, 181, 186, 187, 194, 195, 200, 202, 203, 204, 205, 206, 213, 215, 216, 217, 218, 219, 220, 225, 226, 227, 228, 229, 230, 231, 232, 233, 252
Pliocene, xii, 53, 162, 169, 180, 181, 182
Plio-Pleistocene, 32, 40, 66
pollen, vi, viii, xii, xiii, 31, 32, 43, 55, 57, 65, 193, 194, 195, 196, 197, 198, 199, 201, 202, 203, 204, 206, 207, 208, 209, 210, 211, 212
pollen allergy, viii, xii, 193, 195, 196, 206, 208, 209
pollination, 42, 43, 55
pollinators, 42, 43, 55, 57, 68, 233
pollinosis, vi, viii, xii, 193, 194, 195, 196, 206, 209, 212
pollution, 206, 207
polymer, 118, 119, 120, 123, 126, 130, 135, 147
polysaccharide, 12, 117, 118, 119, 120, 121, 122, 124, 125, 126, 131, 132, 133, 134, 135, 136, 138, 140, 141, 143, 144, 145, 146, 148, 149, 150, 151, 152, 154, 155, 156

Portugal, vii, ix, xiv, 1, 2, 3, 5, 6, 9, 11, 14, 24, 35, 55, 61, 65, 235, 236, 238, 241, 253
prevalence, xiii, 194, 195, 196, 199, 209
Pro j 2, xiii, 194, 203, 205, 206
profilin, xiii, 194, 201, 202, 203, 204, 205, 206, 211, 212
*Prosopis*, xii, 29, 76, 158, 193, 195, 203, 205, 206, 209, 210
*Prosopis juliflora*, 158, 195, 203, 205, 206, 209, 210
protection, xiii, 43, 66, 176, 178, 179, 213, 215, 216, 217
protein family, xiii, 194, 199, 200, 203, 210
proteins, xiii, 166, 194, 197, 198, 199, 201, 203, 204, 205, 209, 216, 217, 219
protons, 118, 119, 123
*Pseudomyrmex*, vi, 42, 43, 166, 177, 182, 184, 186, 213, 219, 220, 221, 222, 224, 225, 226, 227, 229, 230, 233
*Pseudomyrmex veneficus*, 43
*Pseudosenegalia*, 28, 30, 68
pulp, viii, xiv, xv, 3, 18, 23, 235, 236, 237, 238, 239, 240, 241, 242, 243, 244, 246, 247, 248, 249, 250, 251, 253
pulp yield, xv, 4, 18, 236, 237, 238, 239, 240, 241, 242, 250

## Q

quercetin, 13
quinone, 13

## R

rabbits, 45
rain forest, 99, 114
rainfall, 30, 32, 33, 37, 38, 46, 48, 61, 66, 69
rainforests, 31, 33, 34, 35, 61, 180, 217
Raman spectra, 249

Raman spectroscopy, 249, 250
raw materials, xiv, 235, 252
reactivity, xiii, 194, 198, 199, 200, 201, 202, 203, 206, 207, 208, 209, 210, 211, 212
recovery, vii, x, 28, 29, 46, 48, 71, 72, 85
regeneration, v, 41, 46, 60, 66, 83, 84, 85, 93, 96, 97, 98, 102, 103, 107
relationships, xi, 28, 36, 39, 53, 62, 67, 74, 162, 165, 182, 185, 187, 189, 214, 215, 216, 223, 224, 225, 227, 228, 231, 247
respiratory disorders, viii, xii, 193, 195
restoration, vii, x, 64, 84, 85, 95, 97, 98, 101, 102, 108, 110, 111
restoration ecology, v, 83, 85, 103, 108
rhinitis, 196, 208, 209
rhizobia, 55, 57

## S

*S. gilliesii*, 38
*S. praecox*, 38
Sahara, 32, 62
*Salsola kali*, xiii, 194, 200, 203, 204, 205, 206, 210, 211
sapwood, ix, 2, 4, 5, 10, 11, 12, 13, 14, 18, 19, 20, 238, 239
Saudi Arabia, viii, xii, 32, 59, 194, 196, 208
savannas, 31, 32, 37, 38, 41, 49, 50, 53, 60, 62, 68, 69
Schinopsis marginata, 38, 46, 54
SDS-PAGE, 198
seasonal allergic rhinitis, viii, xii, 193, 195
seed, x, 21, 29, 42, 43, 44, 47, 48, 49, 50, 51, 56, 58, 60, 69, 84, 95, 100, 114, 158, 215, 220, 233
seedlings, 48, 51, 60, 68, 85, 86, 93, 98
*Senegalia greggii*, 34
*Senegalia kamerunensis*, 39
*Senegalia mellifera*, 49
*Senegalia senegal*, 57
*Senegalia tenuifolia*, 33

shelter, xiii, 41, 56, 166, 213
shrubs, 37, 40, 46, 49, 65, 67, 69, 166, 173, 181
side chain, xi, 114, 122
signals, 117, 119, 120, 122, 127, 128, 130, 133, 134, 135, 137, 138, 140, 141, 143, 145, 147, 148, 149, 150, 151, 152
skin, 207, 208
smoothness, xv, 236, 243, 247
soil seed bank, x, 84, 95
South Africa, 3, 5, 18, 20, 35, 50, 51, 54, 56, 57, 60, 63, 69, 236, 253
South America, 3, 31, 40, 77, 182, 231
South Asia, 111
Southeast Asia, 108, 111
species diversity, x, 53, 84, 97
species richness, x, 52, 84, 88, 93, 94, 97
spectroscopy, vii, xi, 113, 114, 116, 117, 122, 124, 125, 126, 134, 137, 138, 143, 144, 145, 146, 147, 153, 154, 155, 249, 250
sprouting, 51, 58, 59
symbiosis, xiv, 165, 176, 184, 214, 216, 220, 228
*Syringa vulgaris*, 200, 204, 205

## T

*T. acaciaelongifoliae*, 55
*T. signiventris*, 55
Tanzania, 37, 42
Tasmania, viii, 2, 3, 34, 61, 62
taxa, xiv, 52, 120, 163, 164, 179, 181, 214, 215, 216, 218, 221, 223, 227
taxonomy, 106, 218
temperature, 32, 33, 36, 38, 64, 86, 124, 174, 177, 182, 242
tensile strength, xv, 236, 243, 247
terrestrial ecosystems, 64
tertiary structure, 203, 206
three-dimensional homology, 201

## Index

three-dimensional structures, 199
trade-off, xv, 42, 236
treatment, 19, 43, 62, 115, 171, 249
*Trichilogaster*, 54, 62, 63, 65
Trifolium alexandrium, 36
tropical dry forest, 181
tropical forests, 34, 37, 39, 181, 182

## U

ungulates, 43, 44, 45, 59

## V

*V. aroma*, 38, 43, 47, 49
*V. astringens*, 38, 44
*V. drepanolobium*, 38, 42, 58
*V. seyal*, 42
*V. zanzibarica*, 42, 57
*Vachellia caven*, 34, 47, 53, 55, 56, 59
*Vachellia erioloba*, 36, 38
*Vachellia farnesiana*, xiii, 45, 55, 194, 197
*Vachellia hindsii*, 43
*Vachellia nilotica*, 33, 36
*Vachellia sieberiana*, 48
*Vachellia tortilis*, 32, 33, 37, 39, 44, 67
vegetables, xiii, 194, 201
vegetation, 36, 37, 40, 46, 51, 61, 66, 72, 86, 95, 100, 102, 103, 107, 174, 181, 182
Venezuela, v, vii, xi, 113, 117, 120, 153, 154

vertebrates, 177
vessels, ix, 2, 4, 7, 8, 9, 10
Vietnam, 33, 108, 236
viscosity, 240, 241, 242
Vitis vinifera, 55

## W

water, 10, 36, 37, 38, 39, 41, 42, 43, 45, 46, 53, 54, 57, 58, 62, 95, 99, 116, 129, 166, 173, 219
wildlife, x, 28, 35, 43, 49
wind-pollinated, 195
wood, vii, viii, ix, xiv, xv, 1, 2, 3, 4, 5, 6, 7, 8, 9, 10, 11, 12, 13, 14, 15, 16, 18, 19, 23, 24, 60, 197, 236, 237, 238, 239, 240, 242, 243, 244, 245, 246, 247, 248, 251, 252
wood anatomical characteristics, vii, ix, 2
wood density, vii, ix, 2, 18, 238, 241, 243, 244, 247
wood products, viii, 1, 237
woodland, 61, 68
workers, 18, 177, 197, 208, 221, 222

## Y

yield, viii, xv, 4, 236, 238, 239, 240, 241, 242, 244, 250